AS/A-LEVEL

STUDENT GUIDE

WITHDRAWN

New College Stamford LRC
Drift Road Stamford Lincs.
PE9 1XA
Tel: 01780 484339

AQA

Geography

Component 4: Geographical skills and fieldwork

David Redfern

Hodder Education, an Hachette UK company, Blenheim Court, George Street, Banbury, Oxfordshire OX16 5BH

Orders
Bookpoint Ltd, 130 Milton Park, Abingdon, Oxfordshire OX14 4SB

tel: 01235 827827

fax: 01235 400401

e-mail: education@bookpoint.co.uk

Lines are open 9.00 a.m.–5.00 p.m., Monday to Saturday, with a 24-hour message answering service. You can also order through the Hodder Education website: www.hoddereducation.co.uk

© David Redfern 2017

ISBN 978-1-4718-6417-9

First printed 2017

Impression number 5 4 3 2 1

Year 2021 2020 2019 2018 2017

All rights reserved; no part of this publication may be reproduced, stored in a retrieval system, or transmitted, in any form or by any means, electronic, mechanical, photocopying, recording or otherwise without either the prior written permission of Hodder Education or a licence permitting restricted copying in the United Kingdom issued by the Copyright Licensing Agency Ltd, Saffron House, 6–10 Kirby Street, London EC1N 8TS.

Cover photo: pure-life-pictures/Fotolia

Photograph on p. 95: David Redfern

Typeset by Integra Software Services Pvt Ltd, Pondicherry, India

Printed in Italy

Hachette UK's policy is to use papers that are natural, renewable and recyclable products and made from wood grown in sustainable forests. The logging and manufacturing processes are expected to conform to the environmental regulations of the country of origin.

Contents

Content Guidance

Questions & Answers

Getting the most from this book

Exam tips

Advice on key points in the text to help you learn and recall content, avoid pitfalls, and polish your exam technique in order to boost your grade.

Knowledge checks

Rapid-fire questions throughout the Content Guidance section to check your understanding.

Knowledge check answers

1 Turn to the back of the book for the Knowledge check answers.

Summaries

- Each core topic is rounded off by a bullet-list summary for quick-check reference of what you need to know.

Exam-style questions

Commentary on the questions

Tips on what you need to do to gain full marks, indicated by the icon ⓔ

Sample student answers

Practise the questions, then look at the student answers that follow.

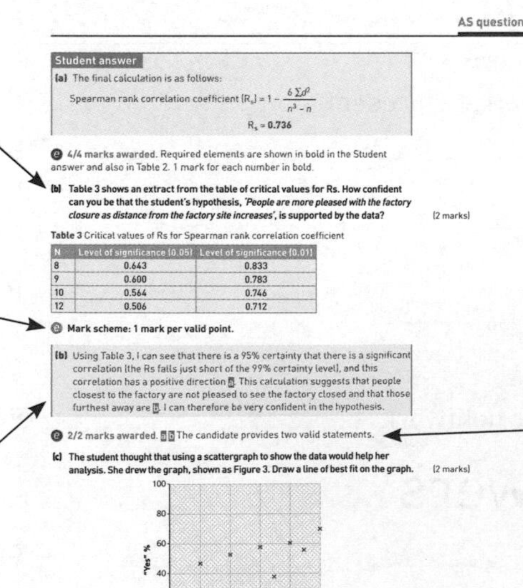

AS questions

Student answer

(a) The final calculation is as follows:

Spearman rank correlation coefficient $(R_s) = 1 - \frac{6\sum d^2}{n^3 - n}$

$R_s = \mathbf{0.736}$

ⓔ 4/4 marks awarded. Required elements are shown in bold in the Student answer and also in Table 2. 1 mark for each number in bold.

(b) Table 3 shows an extract from the table of critical values for Rs. How confident can you be that the student's hypothesis, *'People are more pleased with the factory closure as distance from the factory site increases'*, is supported by the data? (2 marks)

Table 3 Critical values of Rs for Spearman rank correlation coefficient

N	Level of significance (0.05)	Level of significance (0.01)
8	0.643	0.833
9	0.600	0.783
10	0.564	0.746
12	0.506	0.712

ⓔ **Mark scheme: 1 mark per valid point.**

(b) Using Table 3, I can see that there is a 95% certainty that there is a significant correlation (the Rs falls just short of the 99% certainty level), and this correlation has a positive direction a. This calculation suggests that people closest to the factory are not pleased to see the factory closed and that those furthest away are b. I can therefore be very confident in the hypothesis.

ⓔ 2/2 marks awarded. a b The candidate provides two valid statements.

(c) The student thought that using a scattergraph to show the data would help her analysis. She drew the graph, shown as Figure 3. Draw a line of best fit on the graph. (2 marks)

Figure 3 Student's scattergraph

Component 4: Geographical skills and fieldwork 103

Commentary on sample student answers

Read the comments (preceded by the icon ⓔ) showing how many marks each answer would be awarded in the exam and exactly where marks are gained or lost.

About this book

All students of AS and A-level geography following the AQA specification are assessed on the use and application of a range of geographical skills and techniques, and fieldwork. Both of these assessments will be undertaken in the context of the geographical concepts that have been studied, and within a variety of content areas, both physical and human. These concepts and content areas should also provide inspiration for the group and individual fieldwork activities that have to be undertaken during the course.

The **geographical skills and techniques** can be assessed in a variety of ways, including:

(a) in any of the content-based examination questions on any of the AS or A-level papers

(b) in the specific fieldwork and skills questions on Paper 2 (Section B) of the AS examination

(c) in the Non-Examination Assessment (NEA) that each A-level student has to submit

The **fieldwork** is assessed either:

(a) in the specific fieldwork and skills questions on Paper 2 (Section B) of the AS examination

(b) in the NEA that each A-level student has to submit

This guide has three main sections:

Geographical skills: this summarises the specification requirements, and examines each of the required skills and techniques, providing exemplars of most.

Fieldwork: this examines fieldwork at both AS and A-level. In each case, the specification requirements are given, together with advice on how to meet them. At AS, fieldwork and geographical skills are assessed together by means of a written examination. An exemplar assessment is provided in the Questions & Answers section (see below). At A-level, the NEA assesses fieldwork. Guidance is provided on meeting the requirements of the mark scheme for the NEA. This is followed by a number of exemplar fieldwork contexts written in the style of an A-level student. It must be noted that none of these are complete or perfect. Further ideas for individual NEA titles are also provided.

Questions & Answers: this includes some sample fieldwork and geographical skills questions similar in style to those AS students might expect in the exam. There are some sample student responses to these questions as well as detailed analysis, which will give further guidance in relation to what exam markers are looking for in order to award top marks.

The best way to use this book is to be fully aware of what form of assessment/s you are doing, i.e. whether you are doing AS or A-level. Then, you can select the most appropriate aspects of the book.

Content Guidance

Geographical skills

A range of geographical skills and techniques are tested in a number of ways on the AQA specification. They could be assessed in:

(a) examination questions on any of the AS and A-level papers

(b) a more targeted manner in Section B Question 2 of the AS Paper 2

(c) the Non-Examination Assessment (NEA) — the independent fieldwork task for A-level students

The first section states what skills could be assessed, i.e. the specification requirements. The second section follows, with more detail on the application of each of the **specific skills** identified.

Specification requirements

You should develop competence in geographical skills during your study of the course content. You are required to study and make use of a balance of quantitative and qualitative methods across the specification.

During your course you should:

- understand the nature and use of different types of geographical information, including **qualitative** and **quantitative data**, primary and secondary data, images, factual text and discursive/creative material, digital data, numerical and spatial data and other forms of data, including crowdsourced and 'big data'
- collect, analyse and interpret such information, and demonstrate the ability to understand and apply suitable analytical approaches for the different information types
- undertake informed and critical questioning of data sources, analytical methodologies, data reporting and presentation, including the ability to identify sources of error in data and the misuse of data
- communicate and evaluate findings, draw well-evidenced conclusions informed by wider theory and construct extended written argument about geographical matters

Qualitative data Non-numerical data such as photographs and sketches. May also involve the collection of opinions, perspectives and feelings via questionnaires and interviews.

Quantitative data Data in numerical form, often involving measurement, which can often be placed into categories and analysed statistically.

Qualitative skills and quantitative skills

You should develop the following skills with respect to **qualitative data**.

- The use and understanding of a mixture of methodological approaches, including interviews.
- The interpretation and evaluation of a range of source material, including textual and visual sources.

- An understanding of the opportunities and limitations of qualitative techniques, such as coding and sampling, and appreciation of how they actively create particular geographical representations.
- An understanding of the ethical and sociopolitical implications of collecting, studying and representing geographical data about human communities.

You should develop the following with respect to **quantitative data**.

- An understanding of what makes data geographical, and the geospatial technologies (e.g. GIS) that are used to collect, analyse and present geographical data.
- An ability to collect and use digital and geolocated data, and an understanding of a range of approaches to the use and analysis of such data.
- An understanding of the purposes and differences between the following, and the ability to use them in appropriate contexts:
  1. descriptive statistics of central tendency and dispersion
  2. descriptive measures of difference and association, inferential statistics and the foundations of relational statistics
  3. measurement, measurement errors and sampling.
- An understanding of the ethical and sociopolitical implications of collecting, studying and representing geographical data about human communities.

Exam tip

Many of these qualitative and quantitative skills should have been addressed during classroom activities and in your fieldwork. Check back through your notes.

Specific skills

A number of specific qualitative and quantitative skills have to be developed.

- Core skills:
  - the use and annotation of illustrative and visual material, including base maps, sketch maps, OS maps (at a variety of scales), diagrams, graphs, field sketches, photographs, and geospatial, geolocated and digital imagery
  - the use of overlays, both physical and electronic
  - literacy — the use of factual text and discursive/creative material, and coding techniques when analysing text
  - numeracy — the use of number, measure and measurement
  - questionnaire and interview techniques
- Cartographic skills:
  - atlas maps
  - weather maps, including synoptic charts
  - maps with located proportional symbols
  - maps showing movement — flow lines, desire lines and trip lines
  - maps showing spatial patterns — choropleth, isoline and dot maps
- Graphical skills:
  - line graphs — simple, comparative, compound and divergent
  - bar graphs — simple, comparative, compound and divergent
  - scattergraphs, and the use of line of best fit
  - pie charts and proportional divided circles
  - triangular graphs
  - graphs with logarithmic scales
  - dispersion diagrams

- Statistical skills:
  - measures of central tendency — mean, mode, median
  - measures of dispersion — range, inter-quartile range and standard deviation
  - inferential and relational statistical techniques to include Spearman rank correlation and the chi-squared test*, and the application of significance tests
- ICT skills:
  - the use of remotely sensed data
  - the use of electronic databases
  - the use of innovative sources of data, such as crowdsourcing and 'big data'*
  - the use of ICT to generate evidence of many of the skills provided above, such as producing maps, graphs and statistical calculations

* A-level only

Exam tip

Ensure you have made good use of ICT in finding data and processing it, both in class and at home.

The application of skills

Core skills

Labelling and annotation

In an examination and for your fieldwork, you may be asked to label or annotate a photograph (geospatial, geolocated or from digital imagery), sketch, base map, sketch map, diagram or graph. A label is a single word (or short set of words) that simply identifies features shown. At AS and A-level you are more likely to be asked to annotate something. An annotation requires a higher level of labelling, which may be detailed description or offer some explanation, or even some commentary.

Figure 1 shows a field sketch of part of the coastline at Flamborough Head, Yorkshire. It illustrates the difference between a label and an annotation: 'Wave-cut platform' and 'Geo' are examples of labels. 'Steep chalk cliffs, well-jointed rock' and 'Evidence of mass movement — terracettes' are examples of annotations.

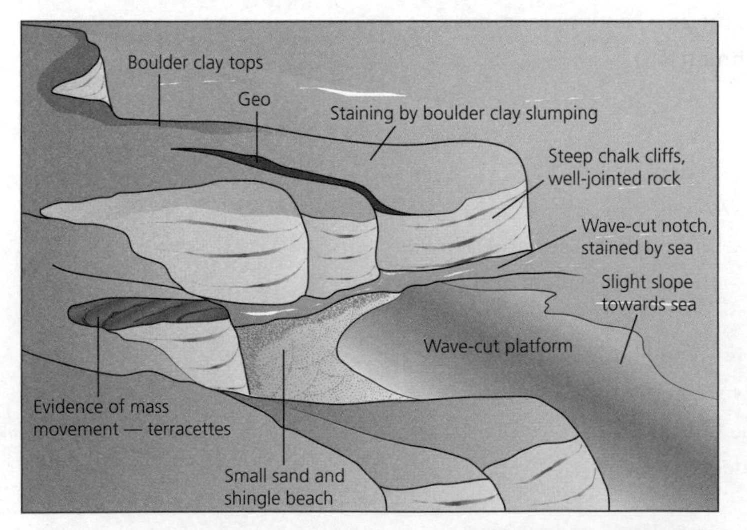

Figure 1 Field sketch of part of the coastline at Flamborough Head, showing the difference between labels and annotations

Ordnance Survey maps

To be able to read an Ordnance Survey (OS) map correctly you need to practise a variety of skills, including:

- grid references (four- and six-figure)
- scale
- compass direction
- height and relief (contour patterns)
- cross section and long section — these are types of simple line graphs that show changes in the shape of the land either across a valley or down a valley. The horizontal (x) axis represents distance and the vertical (y) scale represents height. You should present your section with a realistic scale and make note of the vertical exaggeration if one is present.

It is not necessary to know all the symbols on an OS map off by heart, as in an exam you will be supplied with the key. However, it is quicker and easier if you can learn some of them so that you won't need to constantly refer to the key.

Exam tip

Note that you should be using a range of scales of OS maps during your course.

Sketch maps

Sketch maps are useful for showing the location of a case study or an area of fieldwork. You do not need to include every detail — decide on your priorities. Every sketch map should include the following:

- title
- a scale
- true north arrow
- labels/annotations to indicate important features

Use of overlays

Overlays can be physical, such as tracing paper or a thin plastic sheet, or electronic/digital, such as those on Google Earth or similar online providers. It is likely that you will use physical overlays to identify important points on a photograph, map or sketch. The overlay protects the underlying photo, map or sketch from damage. One example of the use of this technique is to identify the main elements of a transport network in an urban area, such as main roads, railways, canals and cycle routes. Digital overlays allow you to superimpose (or remove) elements of the human landscape, such as roads, urban areas and other features on (or from) a satellite image of a natural landscape.

Literacy skills

Comprehension tasks are a regular feature of examination papers. You are required to read such material at speed and digest its main features. Two command words may feature:

1. 'outline' means identify the key features
2. 'comment on' asks you to infer something from the passage given

You are often expected to use your geographical understanding from your studies and apply it to an unfamiliar, or novel, context. One suggestion that will help you to improve your technique in this area is to read a passage once and then write a précis of the passage to present to a friend or relative.

Exam tip

The assessment of 'novel situations' is a key element of the written examination papers. They often use passages of text combined with a sketch map.

Numeracy skills

It is also important that you hone your numeracy skills, and be aware of number, measure and measurement. For the former, always think about a number in terms of its size in relation to other numbers in the same data set, and also how it compares with average figures for that indicator. Similarly, take note of scales on maps, and make sure to be accurate in your fieldwork measurement.

Exam tip

You may be given a set of data in an examination question that is in two sections, combining both literacy (a passage) and numeracy (a table). The command word used is likely to be 'interpret', which means ascribe meaning. Try to identify links and connections between the two forms of data.

Questionnaire and interview techniques

There is an important difference between questionnaires and interviews, and it is a mistake to use the two terms interchangeably. Questionnaires involve asking a respondent a series of questions, some open, some closed. They typically require written responses although they may be administered face to face. Interviews are one-to-one conversations with a respondent in which the interviewer asks a series of questions that are usually open-ended. In an interview the researcher may ask follow-up questions in order to explore particular areas of interest that arise. Both techniques are used to collect information about attitudes and behaviour that is not available elsewhere as a secondary resource. Questionnaires typically yield quantitative and qualitative data whereas interviews are more likely to generate qualitative information.

Focus groups

This is a qualitative research tool. The researcher brings together a group and asks its members about their perceptions, opinions, beliefs and attitudes towards an idea. This method is widely used in market research, but less often in geographical research, since it can be difficult to set up a representative group.

GIS vs GPS

A geographical information system (GIS) allows us to visualise, retrieve, question, analyse and interpret spatial data in order to understand relationships, patterns and trends. In the 21st century, GIS is essential to understanding what is happening and what will happen in geographic space — planning departments (national and local), utility companies and data research companies all use GIS extensively. There are several free GIS sources of information available on the internet, such as Google Earth and Consumer Data Research Centre (CDRC) maps.

A global positioning system (GPS) provides location information to a receiver anywhere on or near the Earth where there is an unobstructed line of sight to four or more GPS satellites in space. The system provides critical capabilities to military, civil and commercial users around the world. GPS data underpin GIS, especially when providing real-time data such as traffic movement — vehicle satnavs make use of GPS data.

Exam tip

You are likely to use GPS during your fieldwork, for example via the use of a smartphone for location data. GIS should be used in the classroom or at home. Ensure you have done both.

Coding

Coding is the process of organising and sorting qualitative data. Codes serve as a way to label and compile information, which then also allows you to summarise and synthesise what is happening in the data. By linking data collection to the interpretation of the data, coding becomes the basis for developing its analysis.

You can undertake coding in any number of ways, but it usually involves assigning a word, phrase, number or symbol (or even using a set of different-coloured highlighter pens) to each coding category. You could go through all of your textual data (interview transcripts, direct notes, field observations etc.) in a systematic way. You would then code the facts, ideas, concepts and themes to fit the categories you have previously identified. Coding can be extremely complicated, but one example could be to review the transcript of an interview, for example of a survivor of a natural disaster, and use the codes Facts, Opinions, Causes, Impacts (Social, Economic, Environmental) and Responses to organise the data.

Crowdsourcing and 'big data'

Crowdsourcing is the process of obtaining needed services (such as finance), ideas or content by soliciting contributions from a large group of people, and especially from an online community rather than from traditional suppliers.

The use of the social media platform Twitter for geographical information is a good example of crowdsourcing. You can obtain a 'Twitter landscape' by hashtagging (#) a town and seeing what images people from that town have posted to the platform.

'Big data' is a term used to describe the exponential growth and availability of data, both structured and unstructured. Some now qualify this definition by referring to the three Vs of big data: volume, velocity and variety.

Volume

Many factors have contributed to the increase in data volume, for example:

- transaction-based data stored by retailers over time
- unstructured data streaming from social media
- increasing amounts of sensor and machine-to-machine data collection, such as rainfall and river discharge data

In the past, excessive data volume presented storage issues, but with decreasing storage costs and the increasing availability of digital storage systems, other issues emerge. These include how to determine relevance within large data volumes and create valued interpretation from large amounts of relevant data.

Velocity

Today, there is near-constant processing of data.

- Data sets stream in at an unprecedented speed.
- Sensors and smart metering are driving the need to store and deal with torrents of data in near-real time.

Variety

Data today comes in all types of formats, including:

- structured, numeric data in traditional databases, such as the UK census
- unstructured text documents, email, audio and video (via YouTube, Vimeo and similar online outlets)

Exam tip

You are likely to be assessed on coding, crowdsourcing and 'big data' in the skills question on AS Paper 2, and possibly in your NEA fieldwork at A-level. Make sure you have explored each process, albeit briefly.

Cartographic skills

Atlas maps

Make sure that you are able to use an atlas. This means that you should be able to understand and use lines of **latitude** and **longitude** — they are useful when referring to location. Familiarise yourself with using the index in order to quickly locate places. Atlases contain a wealth of maps that can be useful for comparing countries and the collection of secondary data.

Latitude Imaginary lines that run horizontally around the globe, e.g. the Equator.

Longitude Imaginary lines that run vertically around the globe, e.g. the Greenwich Meridian Line.

Synoptic charts

These are weather maps for an area at any one point in time. You are likely to use them during your work on the water cycle. You may also consider using a synoptic chart when completing your fieldwork. The key to the symbols on a synoptic chart will always be given, but you should be aware of the meaning of high and low pressure, and weather fronts (warm and cold).

Exam tip

Do not confuse 'synoptic' in a weather context with the word 'synopticity' — they mean totally different things.

Figure 2 is an example of a synoptic chart for the UK on a day in winter. There is an area of high pressure over NW Scotland.

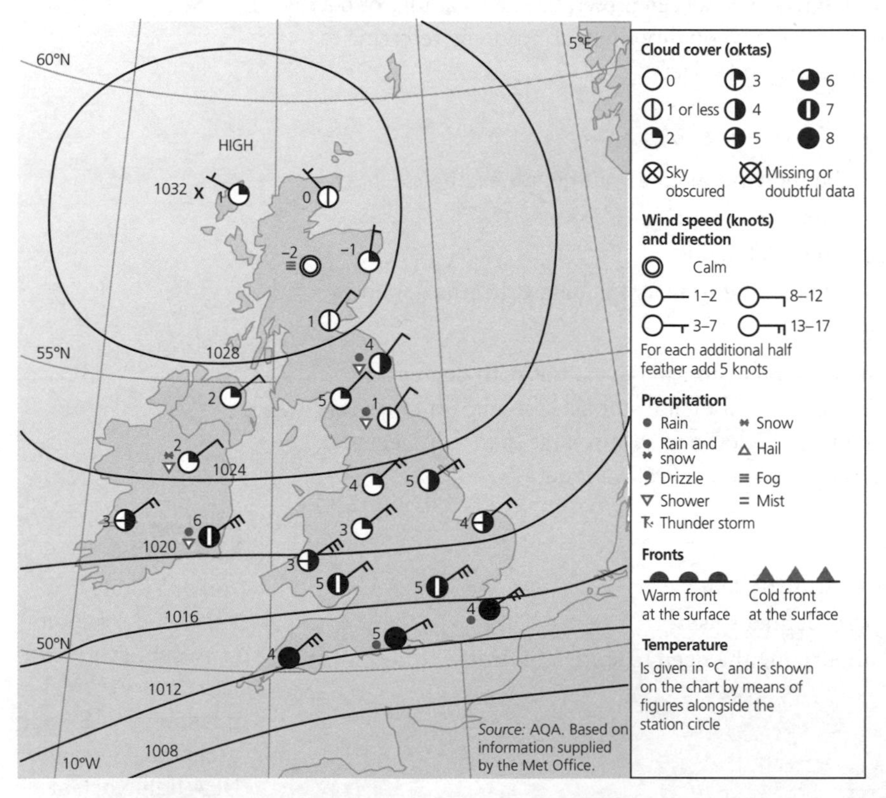

Figure 2 Example of a synoptic chart for the UK on a winter's day

Maps with located proportional symbols

Maps can help you to investigate spatial patterns and compare data between different locations. By using a map with proportional symbols it is possible to investigate spatial patterns together with aspects of volume or size of data. Proportional symbols are symbols that are proportionate in area or volume to the value of data they represent. They can take a variety of forms, including:

- squares
- circles (see Figure 3)
- pie charts
- bar graphs

To make drawing the map worthwhile you need at least three different locations. However, if you have too many it will make the exercise very time consuming. Try to choose a scale such that overlap is avoided, or minimised. Accurate location is important. Other examples of where maps with located proportional symbols are useful include looking at downstream changes in discharge along the long profile of a river, pebble size variations along a stretch of coastline or characteristics of cities/countries/regions around the world.

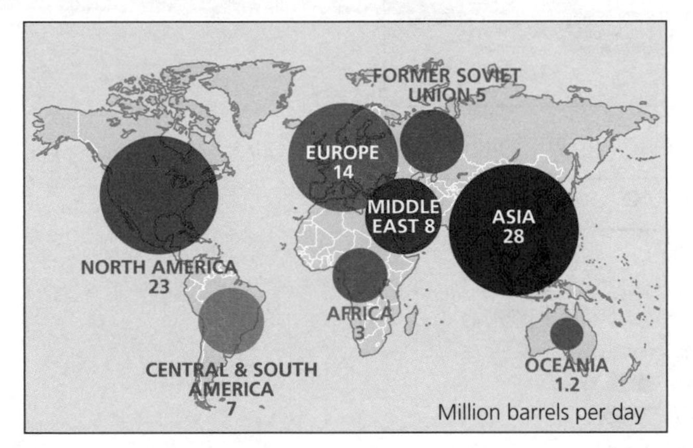

Figure 3 Located proportional circles demonstrating world petroleum consumption by region (1980–2012)

Maps showing movement

Wide arrows, bars or lines show both the direction as well as volume of movement on maps. They are particularly useful in human geography studies.

Flow lines and **desire lines** show the volume of movement and, in both cases, the width of the line is proportional to the quantity of movement. The difference between the two is that a flow line shows the quantity of movement along an actual route whereas a desire line is drawn from the point of origin to the actual destination, and takes no account of the actual route.

Flow line and desire line maps may be useful to show:

- migration routes (see Figure 4)
- movement of traffic across a city (see Figure 5)
- tourist destinations
- origins of visitors, workers or shoppers

Flow line Line on a map showing the volume of movement along a route.

Desire line Line on a map showing the volume of movement from the point of origin to the destination, irrespective of the route.

We draw **trip lines** to show regular trips or journeys that individual people make. For example, we can draw a map to show the places that people commute to from a village location. This technique can be used to work out the catchment areas of shops, schools and other services (see Figure 6).

Trip line Line on a map showing regular trips or journeys that an individual makes.

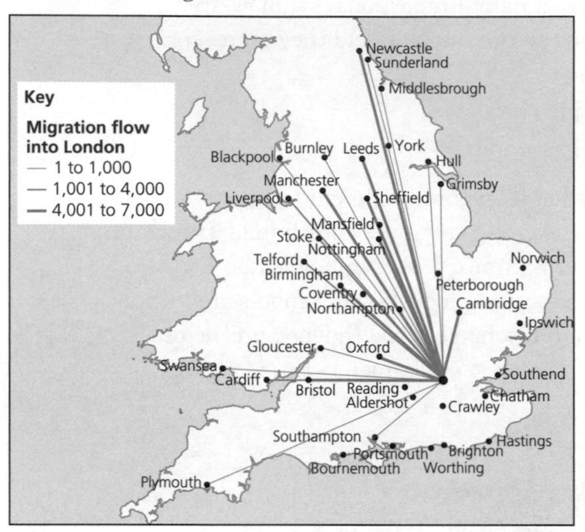

Figure 4 Desire line map showing net migration of people to London (2009–12)

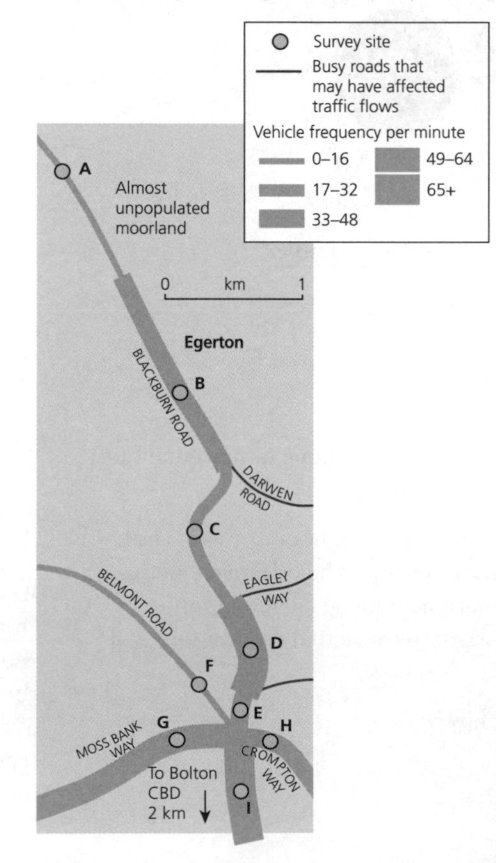

Figure 5 Flow line map showing traffic flows near Bolton

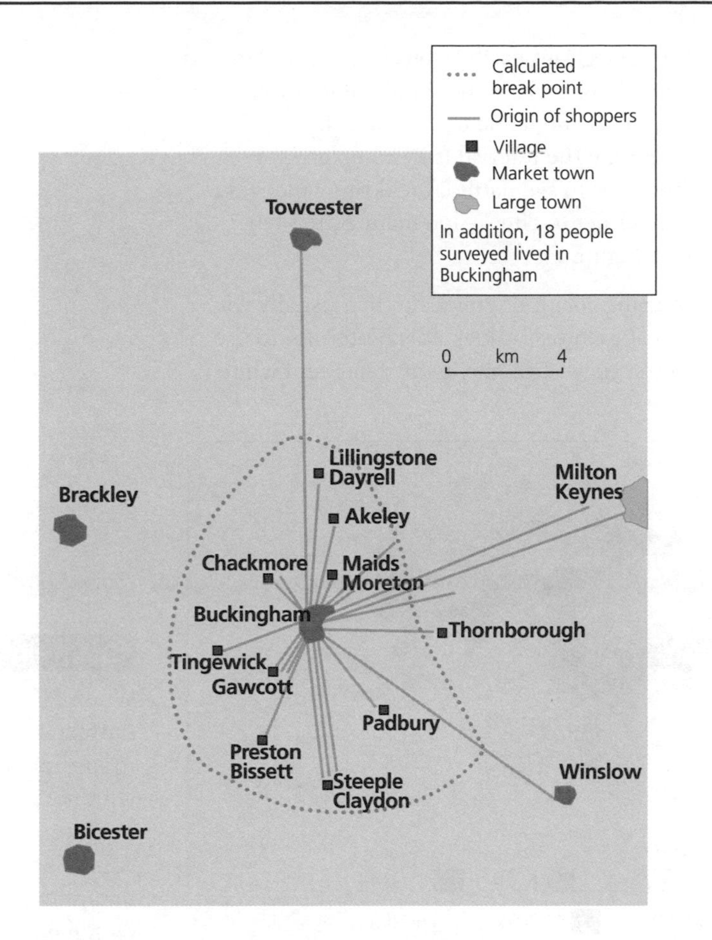

Figure 6 Trip line showing the origins of shoppers on a market day in Buckingham

Maps showing spatial patterns

Choropleth, isoline and dot maps can all be used to show spatial patterns on a map.

Choropleth maps

A **choropleth map** shows the relative density of a characteristic in an area. You complete it by shading in colour, greyscale or line density to show how the data values change from location to location. Choropleth maps are visually striking ways of representing data, as patterns are clearly visible. However, there are also limitations to the technique.

- Data are placed in categories or classes and there may be a large variation within the category/class.
- A choropleth map can take time to construct.
- If you have too few intervals then a large number of locations may have the same shading, making it hard to see a pattern.
- If you have too many intervals it may be difficult to find enough shades of your chosen colour, and again it may be difficult to see a pattern.
- It assumes the whole area under one class of shading has a uniform density. In other words, it doesn't show variations that may occur within an area.

Exam tip

Do not get these three types of map confused. Flow lines show the routes of a group of movements, desire lines the source and destination only, and trip lines only individual movement.

Choropleth map A map showing the relative density of a characteristic in an area through different colour shades or patterns (e.g. population density, per-capita income).

You may be able to identify the intervals or categories easily. However, in practice it is usually more difficult. Make sure that you do not include the same number twice, for example do not use 1–100 and then 100–200. It should be 0–99.9 and then 100–199.99, and so on. One way of determining the interval is by completing a dispersion diagram (see page 25). It is then easy to see natural breaks at which you can place the interval boundaries. Remember not to choose too many or too few categories — ideally, five to six is best (see Figure 7).

Shading can take many forms. If you are using colour or greyscale then usually you shade from dark to light to represent highest to lowest values. It is better not to use the extremes of black and white. Black often suggests a maximum value and white is often used to represent 'no data'.

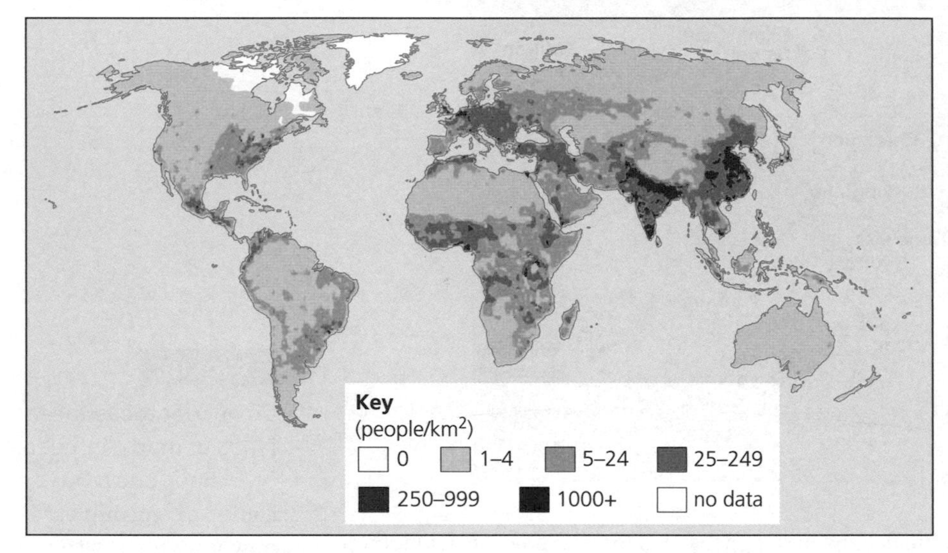

Figure 7 A choropleth map showing global population density estimates (2015)

Exam tip

When describing patterns on choropleth maps, identify areas with high values and areas with low values and then any anomalous situations within each of these areas.

Isoline maps

Isoline maps are maps that show lines drawn on a base map. The lines represent points of equal value. There are many different types and uses of isoline maps, including:

- contour lines on OS maps
- pressure lines on a synoptic chart (known as isobars, see Figure 2)
- temperature (known as isotherms), e.g. to show an urban heat island effect
- travel times for commuters (known as isochrones)
- variations in water depth in a river (see Figure 8)

Isoline map A map of lines on a base map. The lines represent points of same or equal value (e.g. contour lines on OS maps).

Isolines are very useful for examining patterns of distribution. However, a large amount of values are needed, making them more suited to group data collection. The more data points you have, the more accurate your map will be, although it will be more complicated to draw.

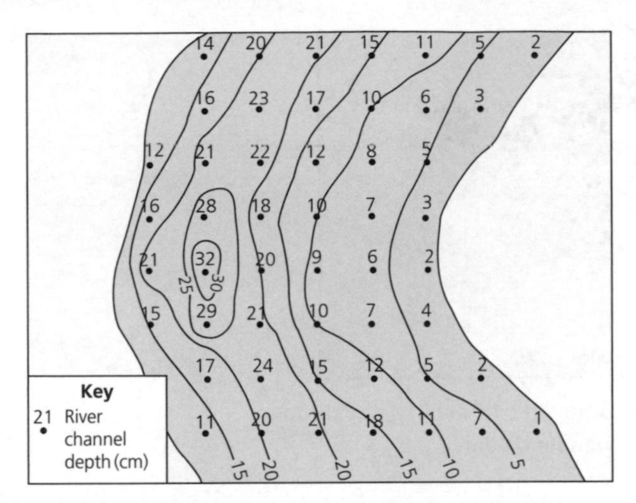

Figure 8 An isoline map showing water depths in a river meander

Dot maps

A **dot map** shows the spatial distribution or density of a variable across an area. Each dot represents the same value and therefore, unlike choropleth maps, it is possible to estimate the numbers in a particular area by counting the dots. A particular advantage of this type of map is that it gives a clear visual impact of patterns in spatial data and especially the locations of any clusters.

Dot map A map that uses dots to show the spatial distribution or density of a variable across an area (e.g. population distribution, incidence of disease).

Some examples of where dot maps are useful include:

- population distribution (see Figure 9)
- distribution of ethnic groups
- incidence of disease
- crime rates

Dot maps are easy to draw but they do have some limitations.

- Where density is very high it is hard to count the dots, making a calculation of actual values difficult.
- Scale is often an issue — some areas may have densities well below the dot value so will appear empty, having a value of 0. For this reason, dot maps can be misleading.
- This type of distribution map can be created more easily today on a GIS system, but they can still be effective as hand-drawn maps.

Knowledge check 1

Having read this cartographic skills section, consider what type/s of map/s you would use for the following scenarios:

- population distribution of a country
- distribution of hospitals in a country
- variations in ethnicity in a city
- origins of customers of a large supermarket
- number of health professionals per 1,000 around the world
- travel times from a major city such as London
- main directions of migrations around the world

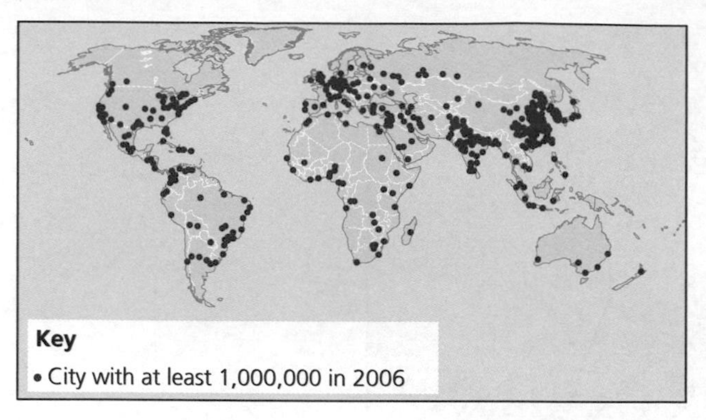

Figure 9 A dot map showing the location of cities with a population of over 1 million people (2006)

Graphical skills

Raw numbers and tables of data (especially large and complex ones) are difficult to understand and interpret. Graphs can tell a story using visual techniques rather than numbers and can help the reader understand the meaning in the data. Another important reason to use a graphical method of presentation is that graphs can highlight patterns and trends in data. Not only does this help you to understand the data, but it also allows you to spot anomalies or irregularities. Graphs can also help you to interpret information at a glance.

General principles

With computer graphics progress, we can now easily produce graphs using spreadsheet software. However, you need to be careful to select the most appropriate type of graph in order to display your particular data — the jazziest graph available may not be the best for your specific data presentation. Note that it is safest to avoid computer-generated three-dimensional graphs, as they can be hard to read and variables may be hidden behind each other.

Good practice with graphs (see Figure 10) includes the following.

- You must include all the key components. Axes must be labelled and the graph must have a title.
- The graph area defines the boundary of all the elements related to the graph including the plot itself and any headings and explanatory text. It emphasises that these elements need to be considered together and that they are separate from the surrounding text.
- The **x-axis** is the horizontal line that defines the base of the plot area. Depending on the type of graph, the *x*-axis represents either different categories (such as years or countries) or different positions along a numerical scale (such as temperature or income).
- The **y-axis** is the vertical line that usually defines the left side of the plot area. If more than one variable is being plotted on the graph then you can have two *y*-axes, one on the left and one on the right. The *y*-axis always has a numerical scale and is used to show values such as counts, frequencies or percentages. For both axes, it is important to choose the right number of categories and labels so that the plot is uncluttered.

x-axis The horizontal axis on a graph, representing different categories or positions along a numerical scale.

y-axis The vertical axis on a graph, representing values.

If the graph you are presenting is based on data from another source other than your own data, you should acknowledge this somewhere within the graph area or title.

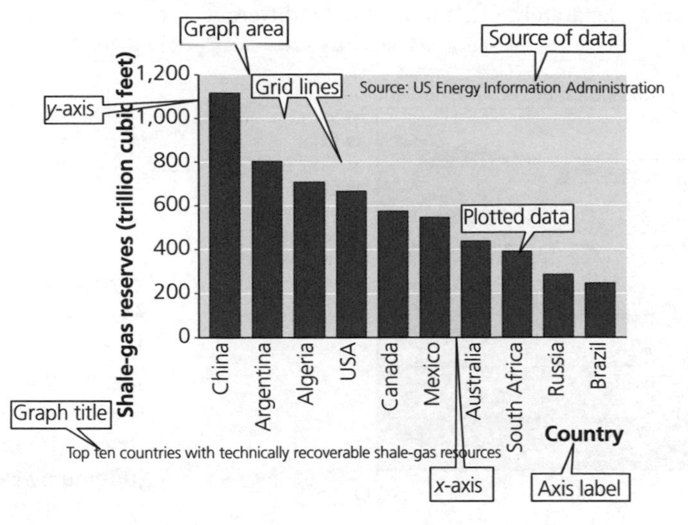

Figure 10 The key components of a graph

Exam tip

As with all forms of data presentation, it is important to be accurate when plotting points on a graph. Inaccuracy will lose (what should be) simple marks.

Line graphs

We usually use line graphs to show time series data (how one or more variable/s change over a continuous period of time). Examples include monthly rainfall or annual unemployment rates. Line graphs are good for identifying patterns and trends in the data. They can also be used for displaying continuous spatial data, for example how pollution levels vary with increasing distance from a source, or how the number of pedestrians changes with increasing distance from a central business district. Line graphs can be **divergent**.

Divergent A graph or chart showing both positive and negative values, usually above and below the *x*-axis.

In a **simple line graph** the *x*-axis represents the continuous variable (for example, year or distance from the initial measurement) while the *y*-axis scale shows units of measurement of the factor being recorded. Several data series can be plotted on the same line graph and this is particularly useful for analysing and comparing the trends in different, but comparable, data sets. These are **comparative graphs** (see Figure 11).

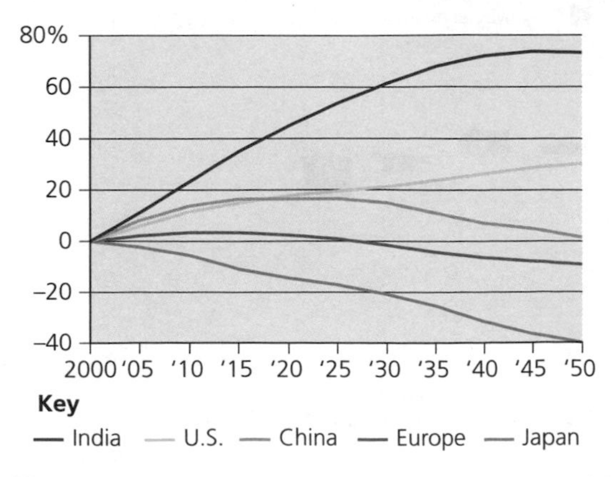

Figure 11 A comparative graph showing changes in working-age populations (actual and projected) (2000–2050)

Sometimes data can be plotted as a **compound line graph**, where categories are placed one on top of the other. Be careful in the interpretation of these types of graph, as you need to read the different areas separately, often indicated by different shading or colours (see Figure 12).

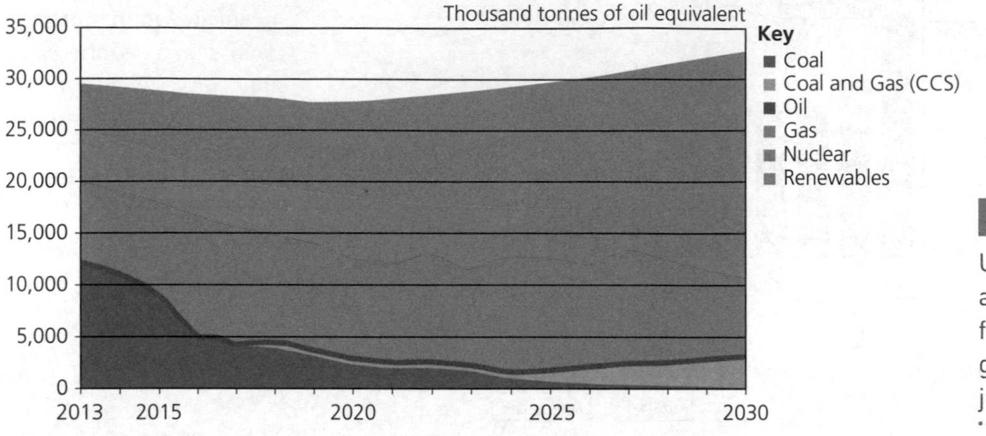

Figure 12 A compound line graph showing future UK sources of energy (2013–30)

Exam tip

Use a ruler to obtain accurate information from a compound graph instead of judging by eye.

Bar graphs

A bar graph, in its simple form, is used to show the differences in frequencies or percentages among discrete categories of data. In a bar graph the categories (*x*-axis) are displayed as rectangles or blocks of equal width, with their height proportional to the frequency or percentage of the data.

Bar graphs are useful for comparing categories of a variable within different groups, for example, a comparison of average wage of adults according to gender. Such a **comparative bar graph** would have two bars, for male and female, and the height of each bar would be scaled to the average wage. Bar graphs can also be divergent (see Figure 13) and **compound**. A common form is a percentage compound graph (see Figure 14).

Compound A graph or chart showing one or more type/s of data, or bars split into sections to show a breakdown of the data.

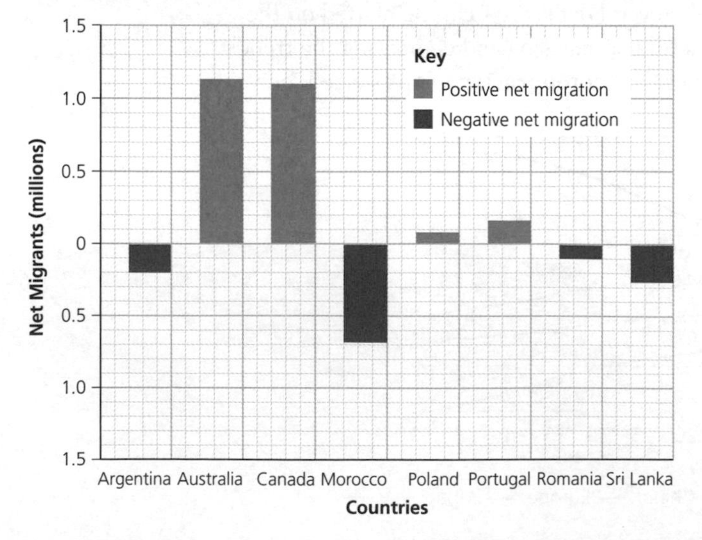

Figure 13 A divergent bar graph showing net migration for selected countries

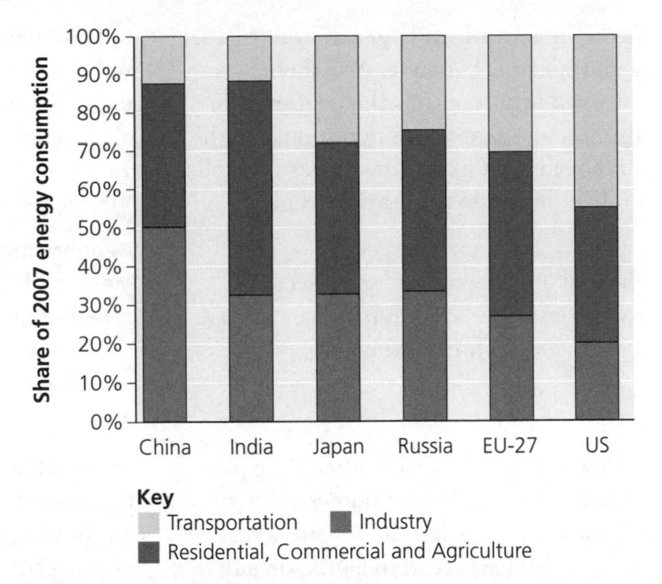

Figure 14 A percentage compound graph showing differences in energy consumption by sector in selected countries and regions (2007)

Scattergraphs

We use scattergraphs to show the relationship between two sets of measurements or data. For example, we may use a scatter plot to present information about the relationship between income equality and life expectancy (see Figure 15). We can add regression lines or **lines of best fit** to a scattergraph. These help the user to understand the strength of the relationship between the two sets of data. Note that the line of best fit does not have to go through the graph's origin. It can also show either a positive or a negative relationship.

Line of best fit A line that goes roughly through the middle point of all the points on a scattergraph — the closer the points to the line, the stronger the correlation.

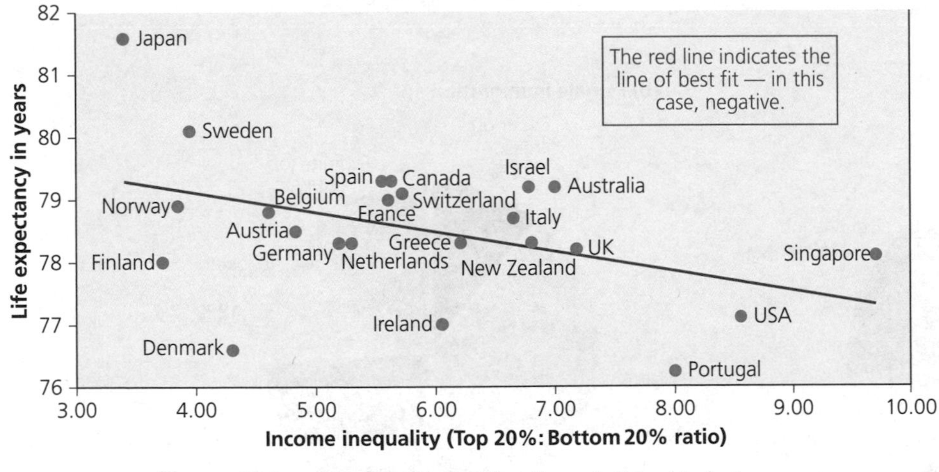

Figure 15 A scattergraph showing the relationship between income inequality and life expectancy

Scattergraphs are often used to work out whether it is worthwhile to carry out further work on the data through hypothesis testing and statistical analysis. You should present the **independent variable** (the quantity you think controls the relationship) on the x-axis and the **dependent variable** (the quantity that varies in response to the independent variable) on the y-axis. So if you thought that plant growth was controlled by the amount of sunlight, you would plot sunlight hours on the x-axis and plant growth rates on the y-axis.

Points lying some distance from the line of best fit are anomalous, and are called **residuals**. For the ease of establishing a possible best-fit relationship, these residuals can be ignored (as long as there are no more than two or three of them).

Independent variable A piece of data that is expected to cause the change in the other variable, presented on the x-axis.

Dependent variable A piece of data that is affected by the change in the other variable, presented on the y-axis.

Residual (anomaly) A point/feature that does not fit the general trend or pattern.

Pie charts

Pie charts are a common technique used to show a frequency distribution. In a pie chart, the frequency or percentage is represented both visually and numerically, so it is quick and easy for a reader to understand the data and what the researcher is conveying (see Figure 16). However, it is true that pie charts tend to be the default option for many students and they are not always the best way to represent data — they can become unwieldy, especially when you have too many categories, while pie charts with fewer than four categories are often better presented as tables. Use them with caution and keep them simple — the overlaying of individual percentages, for example, can add too much clutter and you should use this sparingly.

Proportional divided circles are a number of pie charts drawn at sizes proportional to the total values being represented (see page 13). In this case, the size of the circle is determined by the following formula:

$r = \sqrt{V/\pi}$

Where:

r = radius of the chart

V = the total value to be shown

Exam tip

Note that scattergraphs may show a relationship that has no causal link — the link is just coincidental. In such cases, further research may be needed.

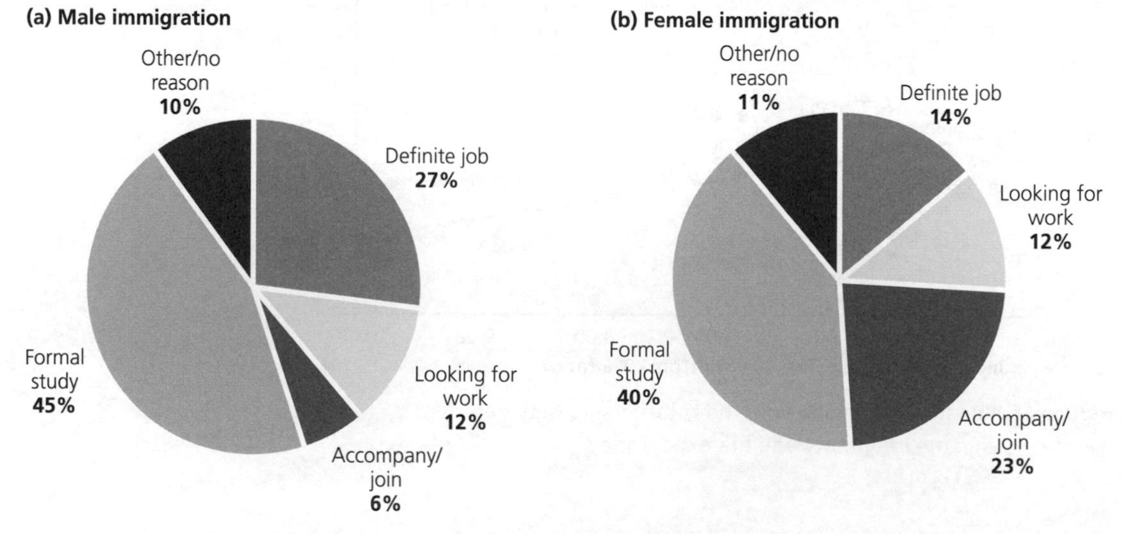

Figure 16 Pie charts showing reasons for migration to the UK by gender (2011)

Triangular graphs

Triangular graphs, as the name suggests, are graphs that are constructed in the form of an equilateral triangle (see Figure 17). They can only be used if the conditions below apply, and for these reasons they are fairly limited in their use.

- The data must be able to be divided into three component parts.
- The data must be in the form of percentages.
- The percentages must total 100.

Each side of the triangle represents one axis and one component, and measures from 0% to 100%. From each axis, lines are drawn at 60-degree angles to carry the values.

Triangular graphs are useful because you can show a large amount of data on one graph. As with pie charts, it is easy to see relative proportions and identify the dominant variable. However, triangular graphs can be difficult to interpret and care must be taken not to make errors reading off the incorrect axis. On the other hand, they have an in-built checking system, as all values for one plot must total 100. If the total is not 100, an error has been made.

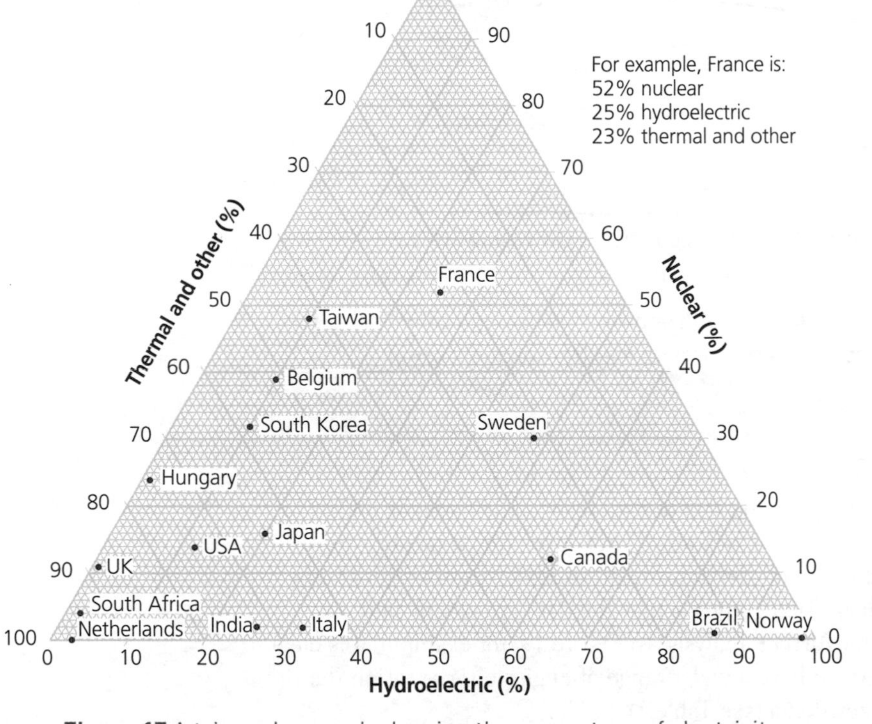

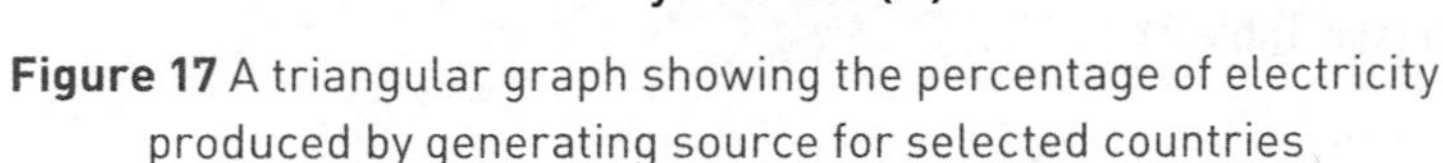

Figure 17 A triangular graph showing the percentage of electricity produced by generating source for selected countries

Graphs with logarithmic scales

Logarithmic graphs are a useful form of line graph where a large range of data has to be plotted. In an arithmetic line graph the scale increases by equal amounts. **Logarithmic scales** differ from this in that the scales are divided into a number

> **Exam tip**
>
> When constructing a pie graph, it is good practice to begin the sequence of segments at 12 o'clock, work clockwise and maintain the same sequence of segments to make comparison easier.

> **Exam tip**
>
> One way to read a triangular graph is to imagine it as three separate pyramids, one for each variable. Each side acts as a base for that variable (0%) and each apex represents 100%.

Logarithmic scale A scale that is divided into cycles or intervals, which increase tenfold each time.

of cycles, each representing a tenfold increase. An example of this principle is: the first cycle ranges 1–10, the second cycle 10–100, the third 100–1,000, and so on. When both the *x*- and *y*-axis scales are logarithmic they are called log–log graphs (or log–log plots). If only one axis is logarithmic, it is known as a semi-logarithmic graph.

One disadvantage of these graphs is that you cannot use positive and negative values on the same graph. Figure 18 shows an example of semi-logarithmic graph paper, where the logarithmic scale is shown on the vertical *y*-axis. Examples of the use of logarithmic scales include the moment magnitude scale, used to measure earthquakes, and the Hjulström curve, used in river studies.

Exam tip

Note that logarithmic graphs are useful when interpreting rates of change — the steeper the line, the faster the growth.

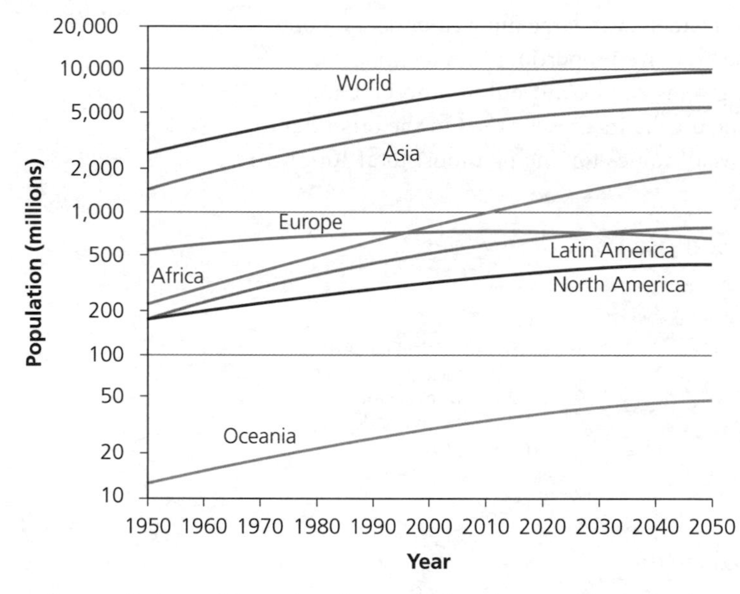

Figure 18 A semi-logarithmic graph showing world population growth and projected population growth in different continental regions (1950–2050)

Dispersion diagrams

Dispersion graphs are useful for comparing sets of data (especially patterns both within and between sets) for two or more locations. There are numerous occasions when dispersion graphs should be used, and they are often under-used by geography students. A key advantage of dispersion graphs is that the data are clearly laid out along a vertical axis as a basis for further analysis. In Figure 19, for example, it is clear that the corries on the Isle of Arran have a wider range of angles of orientation than the corries in the Glyders of Snowdonia (see Table 1).

Table 1 The lip orientation of 15 corries in the Glyders (Snowdonia) and 15 on the Isle of Arran (°)

Corrie number	1	2	3	4	5	6	7	8	9	10	11	12	13	14	15
Glyders	30	60	45	55	75	50	80	50	10	15	10	35	45	50	85
Arran	5	5	10	55	15	30	95	5	185	70	120	40	30	115	110

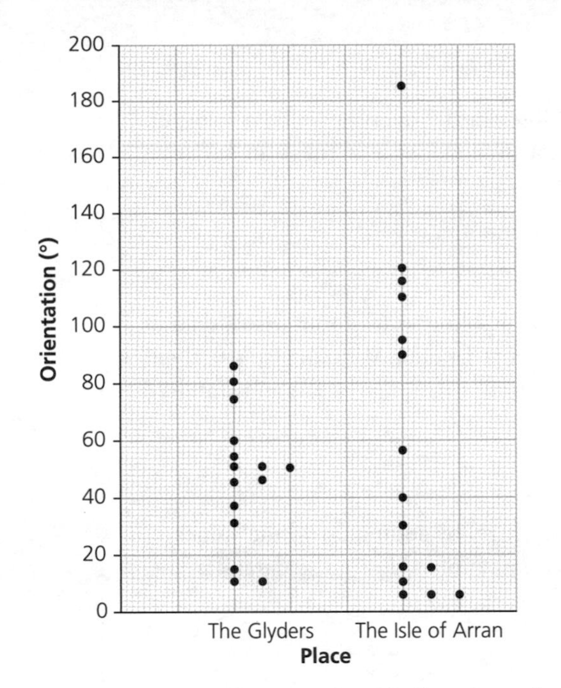

Figure 19 Dispersion diagram of corrie lip orientation for the Glyders (Snowdonia) and the Isle of Arran

Knowledge check 2

Having read this graphical skills section, consider what type/s of graph/s you would use for the following scenarios:

- the changing length of a glacier/sand spit/alluvial fan over time
- the percentage of global deaths from a range of non-communicable diseases
- the energy mix of a country
- the top ten countries with the highest urban populations
- the potential relationship between air pollution levels and the number of dementia cases

Statistical skills

For the following section, refer to the data provided in Table 2.

Table 2 A river in northern England: discharge in 4 months in cumecs (1994–2012)

Year	January	April	July	October
1994	56.68	15.23	14.04	72.61
1995	70.96	6.35	6.15	42.59
1996	86.42	25.85	6.55	72.93
1997	77.86	7.50	1.88	67.91
1998	24.45	45.05	40.87	30.57
1999	69.94	24.68	6.03	45.62
2000	33.26	27.11	31.89	70.23
2001	81.19	17.29	42.80	47.27
2002	34.51	31.16	3.60	42.68
2003	88.81	13.95	18.01	50.07
2004	49.33	34.81	10.48	33.12
2005	31.69	31.58	6.16	27.56
2006	79.70	42.05	21.00	11.98
2007	79.46	50.62	5.16	31.17
2008	93.93	12.78	7.17	25.22
2009	21.16	16.78	9.56	41.18
2010	10.00	10.00	3.60	10.20
2011	29.98	17.58	31.00	80.60
2012	99.70	32.60	32.05	46.65

Measures of central tendency

Measuring central tendency refers to ascertaining a measure of the 'middle' value of a data set. There are three ways of measuring central tendency, and these are:

1 mean
2 mode
3 median

These techniques are very useful to geographers, enabling us to summarise a data set by giving the mid-value or most frequently occurring data. We can also use them as part of more complex techniques, such as inter-quartile range (IQR) and standard deviation.

Mean

We calculate the mean (sometimes called the average) by adding up all the values in a data set and dividing the total sum by the number of values in the data set. The formula for calculating the mean is as follows:

$$\bar{X} = \frac{\Sigma X}{n}$$

The mean is particularly useful if the data have a small range. However, if the range is large, the extreme values will heavily influence the mean and could give a distorted picture.

Referring to Table 2, the mean discharges for the 4 months are:

- January: 58.90 cumecs
- April: 24.37 cumecs
- July: 15.68 cumecs
- October: 44.75 cumecs

Mode

This is the value that occurs most frequently in a set of data. You need to know all values before calculating the mode. Mode is of no use if there are no repeating values. There may be more than one mode — this is called 'bi-modal'. The mode is often useful when classifying data into groups or classes. It is useful to see which classification occurs most frequently. This is called 'modal class'.

There are no repeating values in Table 2, so no modes can be given.

Median

This is the middle value in a data set. The data need to be placed in rank order before you can calculate the median. If there is an odd number of values, perform the following calculation to work out the median value:

$$\frac{n+1}{2}$$

Where:

n = number of values in the data set

For example, if you have 23 values in a data set, the median will be the 12th value in the rank order. If the number of values is even, the median is the mean of the middle two values. So if there are 24 values, add the values for the 12th and 13th positions and divide by two.

The median value often needs to be supported by other techniques such as inter-quartile range. However, unlike the mean, it not affected by extreme values.

Referring to Table 2, which has 19 sets of data for each month, the median discharges for the 4 months are the tenth item of data (when ranked highest to lowest):

- January: 69.94 cumecs
- April: 24.68 cumecs
- July: 9.56 cumecs
- October: 42.68 cumecs

None of these measures gives an accurate picture of the distribution of data in Table 2. On their own they are of limited value. However, it can be seen from the above example that some limited judgements can still be made. In two of the measures the central tendency for river discharge is largest in January and lowest in July. The spread of data is, however, quite large for January, and the extreme values of 10.00 and 99.70 may be making the mean value higher. The uneven spread of data for January means that this data set is 'skewed'. In general, the greater the skew in a data set, the greater variation in the three measures of central tendency.

To improve the usefulness of the above calculations, you should next calculate measures of the dispersion or variability of the data.

Exam tip

Be clear on the differences between the three indices of central tendency (mean, mode, median) and of the different ways in which they are calculated.

Measures of dispersion

Measures of dispersion are techniques used to measure the spread of data. Range, inter-quartile range and standard deviation allow you to analyse your data in more depth, and to examine how the data is spread around either the mean or the median.

Range

This is simply the difference between the highest value and the lowest value. It gives you a basic idea of the spread of data, but like the mean, it is affected by extreme values. Therefore, an anomaly can give a false picture. Referring to Table 2, the ranges are as follows:

- January: 89.70
- April: 44.27
- July: 40.92
- October: 70.40

Therefore we can see that January has the largest range and July the smallest.

Inter-quartile range

The inter-quartile range (IQR) is a measure of dispersion around the median. We calculate it by ranking the data (highest to lowest) and placing the data into quarters, separated by quartiles, on a dispersion diagram (as shown earlier, in Figure 19). The top 25% of the data is placed above the upper quartile (UQ) and the bottom 25% is placed below the lower quartile (LQ). The IQR is the difference between the 25% and 75% quartiles, or UQ minus LQ (see Figure 20).

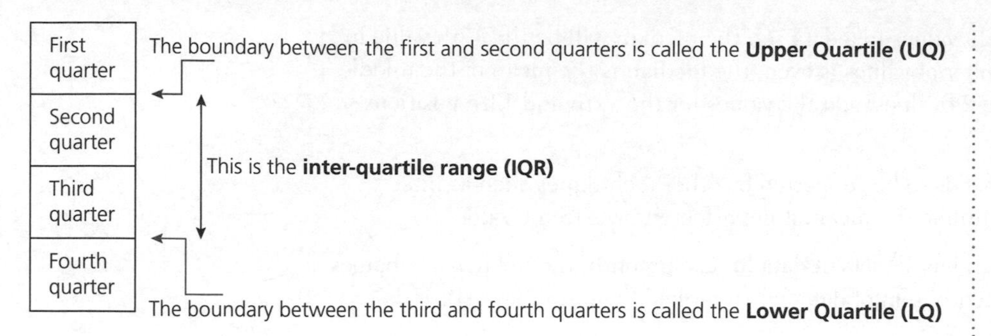

Figure 20 The inter-quartile range

The IQR is more useful than the range in indicating the spread of data, as it takes away any extreme values (i.e. those occurring in the 1st Quarter and 4th Quarter) and considers the spread of the middle 50% of the data around the median value. There are a series of formulae to work out the IQR:

Upper Quartile (UQ) = $\frac{n+1}{4}$ th position in the data set (ranked from highest to lowest)

Lower Quartile (LQ) = $\frac{(n+1) \times 3}{4}$ th position in the data set (ranked from highest to lowest)

Inter-quartile range (IQR) = UQ – LQ

Using the data in Table 2 as our starting point, we can produce the IQR data, as demonstrated in Table 3:

Table 3 IQR data, using Table 2 as a starting point

	January	April	July	October
UQ	81.19	32.60	31	67.91
LQ	31.69	13.95	6.03	30.57
IQR	49.50	18.65	24.97	37.34

There is least variation in April and greatest in January.

Exam tip

The subject of geography has long had the convention of ranking data from highest to lowest. You will find it more straightforward to do so.

Standard deviation

Standard deviation (SD) is a measure of the degree of dispersion about the mean. The formula for calculating standard deviation is as follows:

$$\sigma = \sqrt{\frac{\Sigma(x-\bar{x})^2}{n}}$$

Where:

σ = standard deviation

Σ = sum of

$\bar{x}$ = mean

n = number in the sample

It is possible that two sets of data could have the same mean but have a very different spread of data. SD will tell you the extent of this spread — in other words, how reliable the mean is. A low SD indicates that the data points tend to be very close to the mean, whereas a high SD indicates that the data is spread out over a larger range of values and the mean is less reliable, as there is obviously a lot of variation in the sample. Hence, SD is used when you want to compare the dispersion of two or more sets of data. An important aspect of SD is that it links the data set to the **normal distribution** (see Figure 21).

Normal distribution A theoretical frequency distribution that is symmetrical about the mean.

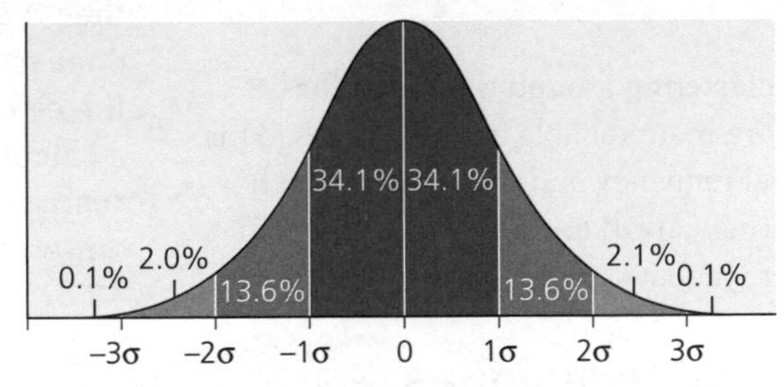

Figure 21 The normal distribution (sometimes known as the bell-shaped curve)

In a normal distribution:

- 68% of the values lie within ±1 standard deviation of the mean
- 95% of the values lie within ±2 standard deviations of the mean
- 99% of the values lie within ±3 standard deviations of the mean

Exam tip

The link between standard deviation and the normal distribution allows you interpret the outcome more easily. Try to remember these percentages.

Worked example of standard deviation

Refer back to our data in Table 2. In Table 4, the calculation for standard deviation has been calculated for the river discharge in January.

Table 4 Standard deviation calculation for the river discharge in January

January discharge	$x - \bar{x}$	$(x - \bar{x})^2$	
56.68	−2.22	4.93	
70.96	12.06	145.44	
86.42	27.52	757.35	
77.86	18.96	359.48	
24.45	−34.45	1,186.80	
69.94	11.04	121.88	
33.26	−25.64	657.41	
81.19	22.29	496.84	
34.51	−24.39	594.87	
88.81	29.91	894.61	
49.33	−9.57	91.58	
31.69	−27.21	740.38	
79.90	21	441	
79.46	20.56	422.71	
93.93	35.03	1,227.10	
21.16	−37.74	1,424.31	
10.00	−48.90	2,391.21	
29.98	−28.92	836.37	
99.70	40.80	1,664.64	
$\Sigma x = 394$	$\bar{x} = 58.90$	$\Sigma (x - \bar{x})^2 = 14{,}458.91$	$\frac{\Sigma (x - \bar{x})^2}{n} = \frac{14{,}458.91}{19}$
			$\sigma = 27.59$

Standard deviation for the river discharge in January = 27.59

The other SD calculations are:

- April: 12.59
- July: 13.10
- October: 19.96

This exercise in SD suggests that there is more clustering around the mean in the spring and summer months. The mean is therefore more reliable at these times. This is also supported by the other measures of central tendency and dispersion, which have indicated that there is less dispersion in spring (April) and summer (July) and more dispersion in autumn (October) and winter (January).

Exam tip

If you can use an Excel spreadsheet, then the calculation is done for you. If you do the calculation using a calculator, don't rush it — you'll avoid any errors.

Inferential and relational statistical techniques

Comparisons are often made between two sets of data to see if there is a relationship between them. However, we should always recognise that the existence of a statistical relationship does not prove a causal link. Further investigation is needed to establish whether or not there is a causal link — the statistical evidence may just suggest there could be one.

The Spearman rank correlation test

One way of showing a possible relationship between two sets of data is a scattergraph (see Figure 15). Spearman rank is another way of testing a relationship. When you draw a scattergraph you can see by eye if there is a relationship, but you will probably not be able to clearly assess the strength of the relationship, as many points may be some distance from the line of best fit. We therefore use Spearman rank correlation (see below Table 5) to test the strength of the relationship between two sets of data, providing you with a numerical value between 0 and +1, or between 0 and –1. Once you have this figure you can then test its significance — this means the likelihood of your results occurring by chance.

The test can be used with any set of raw data or percentages but it is only suitable if all of the following criteria apply.

- You have two data sets, which you believe may or may not be related.
- You're using at least ten pairs of data.
- You're using no more than 30 pairs of data (as this makes the exercise unwieldy).

Procedurally, you should start by stating your hypothesis, such as *'As* x *increases then so does* y*'*. You should then be prepared to accept the opposite — *'There is no relationship between* x *and* y*'*. This is called the **null hypothesis**.

Null hypothesis Usually that 'there is no relationship between variable 1 and variable 2'.

A worked example is given below. Once you have completed the table and have your result at the end of the calculation you should have a figure between –1 and +1. This indicates the strength and type of relationship.

- A result close to +1 indicates a positive relationship (i.e. as one set of data increases so does the other).
- A result close to –1 indicates a negative relationship (i.e. as one set of data increases the other decreases).
- A result close to 0 means that there is no relationship and you must accept the null hypothesis.

Exam tip

In an examination you will not be expected to learn the formula for Spearman rank or any other statistical test.

The Spearman rank correlation has a number of **strengths**.

- It gives an outcome that is objective.
- It enables you to demonstrate a numerical relationship between two sets of data, although it must be stressed again that this may not be a causal relationship.
- You can state whether the relationship is significant or if your results were just a fluke.
- It is less sensitive to anomalies in data as each piece of data is ranked — large differences could only be one rank apart.

The Spearman rank correlation also has a number of **weaknesses**.

- It does not tell you whether there is a causal link (i.e. that change in one variable leads to a change in another), but rather simply that a relationship exists.
- Too many 'tied ranks' can affect the validity of the test.
- It could be subject to human error, such as inaccurate calculations.

Worked example of a Spearman rank correlation

You are investigating the relationship between the percentage of unskilled workers and the percentage unemployed to ascertain the variations in social and economic conditions in an urban place you are studying.

The null hypothesis is '*that there is no relationship between the percentage of unskilled workers and the percentage unemployed in an area*'.

Using your data, you construct Table 5.

Table 5 Variations in social and economic conditions in an urban area in England

Ward	A % unskilled workers	Rank (A)	B % unemployed	Rank (B)	Difference in rank (A – B) (d)	d^2
1	5.5	13	9.4	15.5	2.5	6.25
2	6.6	10	15.8	6	4	16
3	6.0	11	11.7	13	2	4
4	2.2	17	8.9	17	0	0
5	15.6	1	23.0	1	0	0
6	8.4	6	18.9	3	3	9
7	7.3	8	13.9	10	2	4
8	8.5	5	15.1	8	3	9
9	8.9	4	14.2	9	5	25
10	11.6	3	17.5	4	1	1
11	12.5	2	22.0	2	0	0
12	5.6	12	12.4	11	1	1
13	8.1	7	15.4	7	0	0
14	1.8	18	6.2	18	0	0
15	7.2	9	16.7	5	4	16
16	4.3	15	12.3	12	3	9
17	4.2	16	11.6	14	2	4
18	5.1	14	9.4	15.5	1.5	2.25
						$\sum d^2 = 106.5$

Exam tip

As elsewhere, do not rush your work here, in order to avoid errors.

Spearman rank correlation coefficient

$$(\mathbf{R_s}) = 1 - \frac{6\,\Sigma d^2}{n^3 - n} = 0.890$$

Note that Wards 1 and 18 have tied ranks for percentage unemployed.

You now need to do another test to check whether your result could have occurred by chance — in other words, 'how significant is your result'? To do this you have to compare your result with a table of **critical values** (see Table 6). First, look at the number of pairs of data you have — in this case there are 18. You then need to decide which **significance** level you are going to use. For geographical purposes, you would usually use the 0.05 significance level. This means that there is a 5 in 100 chance of the results occurring by chance. Or to put it another way, if other researchers completed the same experiment, 95 out of 100 would get the same result — therefore there is a strong chance of a relationship between the two items. You then need to see whether your Rs result is *above* the critical value for the number of pairs you have. If your Rs value is *below* the critical value, you must accept the null hypothesis, i.e. you cannot be sure that your relationship is significant.

Critical value Calculation used with many statistical techniques to test the significance (or confidence) level. Each technique will have a table of critical values against which the result of the statistical test is compared.

Significance The degree to which you can be confident that your results did not occur by chance. Usually in geography, we use the 95% (0.05) and 99% (0.01) confidence levels.

Table 6 Critical values for Rs

n	0.05 significance level	0.01 significance level
10	+/- 0.564	+/- 0.746
12	0.506	0.712
14	0.456	0.645
16	0.425	0.601
18	0.399	0.564
20	0.377	0.534
22	0.359	0.508
24	0.343	0.485
26	0.329	0.465
28	0.317	0.448
30	0.306	0.432

We can see that in our example the Rs value of 0.890 is well above the 0.05 significance level of +0.399. It is also well above the 0.01 significance level of +0.564. This means that there is a very low (1 in 100) chance of the results occurring by chance, and we can reject the null hypothesis. Having rejected the null hypothesis, you can accept that there is a strong relationship between the percentage of unskilled workers and the percentage unemployed, and that it is highly significant.

Exam tip

Be clear in your understanding of the purpose of significance testing.

The chi-squared test

We use the chi-squared test to investigate spatial distributions. It looks at frequencies or the distribution of data that you can put into categories, for example pebble shapes at different sites along a beach or frequencies of plant types at different stages of a vegetation succession. Chi-squared is a comparative test, as it compares actual data collected against a theoretical random distribution of the data.

The data collected is called the **observed data (O)**. The theoretical, random, distribution is called the **expected data (E)**.

For a chi-squared test to be conducted, the following must occur.

- The data need to be organised into categories.
- The data cannot be in the form of percentages and must be displayed as frequencies.
- The **total** amount of observed data must exceed 20.
- The **expected** data for each category must exceed 4.

As with other statistical tests, chi-squared requires and tests a null hypothesis. The null hypothesis is that *'there is no significant difference between the observed distribution and the expected distribution'*.

The strength of chi-squared lies in the fact that, as with other statistical tests, you can check the significance of your results. Also, as with other statistical tests, a weakness includes human error in calculating x^2. It also does not explain why there is, or there is not, a pattern to the distribution. As with Spearman rank, this will need further investigation.

Chi-squared (x^2) is calculated using the following formula:

$$x^2 = \Sigma \frac{(O-E)^2}{E}$$

Exam tip

In an examination you will not be expected to learn the formula for chi-squared or any other statistical test.

Worked example of a chi-squared test

A group of students investigated the orientation of pebbles in an exposed bed of glacial till. The glacial till was situated near the lip of a corrie in the Lake District. The students wanted to investigate whether there was a pattern to the orientation of the long-axis of the till. Their hypothesis was: *'There is a significant trend in the orientation of pebbles within the glacial till'*.

They measured the orientation of 40 pebbles and placed their results into four categories:

1. 0–45° = 2 pebbles
2. 46–90° = 10 pebbles
3. 91–135° = 23 pebbles
4. 136–180° = 5 pebbles

The data suggests that there is a preferential direction, but as this is maybe due to chance, the students carry out a chi-squared test. The test begins with the null hypothesis that *'there is no significant difference between the observed orientation of pebbles and an expected random orientation'*.

Next the students created a table (see Table 7).

Table 7 Chi-squared value of the orientation of pebbles in an exposed bed of glacial till

			A	B	C
Orientation	Observed (O)	Expected (E)	O − E	(O − E)^2	$\frac{(O - E)^2}{E}$
0–45°	2	10	−8	64	6.4
46–90°	10	10	0	0	0
91–135°	23	10	13	169	16.9
136–180°	5	10	−5	25	2.5
					x^2 = 25.8

The chi-squared value x^2 = 25.8

The result by itself is meaningless. The students needed to test its significance. To do this, they worked out the degrees of freedom using the formula $(n - 1)$, where n is the number of observations — in this case the number of categories, which contained observed data. Therefore for this example, $n = 4$, so the **degrees of freedom** are **4 − 1 = 3**.

Using Table 8, they compared their result for x^2 at the degree of freedom of 3 for the 0.05 and 0.01 significance levels. If the x^2 result is the *same or greater* than the value given in the table, the null hypothesis can be rejected.

Table 8 Critical values of chi-squared

Degrees of freedom	Significance level 0.05	Significance level 0.01
1	3.84	6.64
2	5.99	9.21
3	7.82	11.34
4	9.49	13.28
5	11.07	15.09
6	12.59	16.81
7	14.07	18.48
8	15.51	20.09
9	16.92	21.67
10	18.31	23.21
11	19.68	24.72
12	21.03	26.22
13	22.36	27.69
14	23.68	29.14
15	25.00	30.58

They wrote a summary statement to express the result for the chi-squared test:

'At 3 degrees of freedom, the x^2 result of 25.8 is above the 0.01 critical value of 11.34. Therefore we can reject the null hypothesis and accept that the orientation of the till did not occur by chance and is not randomly orientated. There is therefore a significant trend in the orientation of the pebbles.'

Significance The degree to which you can be confident that your results did not occur by chance. Usually the 95% (0.05) and 99% (0.01) levels are used.

Degrees of freedom In statistics, the number of values in the final valuation of a statistic that are free to vary.

Exam tip

Be clear in your understanding of the purpose of significance testing.

Critical value Calculation used with many statistical techniques to test the significance (or confidence) level. Each technique will have a table of critical values against which the result of the statistical test is compared.

Summary

After studying this section, you should have an:

- awareness of the specification requirements of the geographical skills that need to be understood at both AS and A-level, known as 'the skills checklist'
- understanding of how to use and apply the core and other general skills that are needed when undertaking any skills-based activity
- awareness of the need to make use of ICT skills when collecting and processing data
- understanding of how to use and apply a range of cartographic skills in a variety of contexts
- understanding of how to use and apply a range of graphical skills in a variety of contexts
- understanding of how to use, apply and interpret a range of statistical skills in a range of contexts

Fieldwork

Fieldwork at AS

Specification requirements

You must undertake the equivalent of at least **2 days** of geographical fieldwork. These days must be based on both physical geography fieldwork and human geography fieldwork, and the themes for investigation should be derived from the AS specification. It is likely that schools and colleges will plan 1 day focusing largely on physical geography and another day focusing predominantly on human geography. However, there is no requirement that these two areas of study have to be equal in terms of emphasis or time allocation.

Students will probably visit contrasting environments to undertake the necessary physical and human geography fieldwork, but the locations need not be far apart. For example, looking at beach processes and then going into a nearby town to study levels of deprivation and/or environmental quality in urban areas would satisfy the requirements.

There are no levels of control applicable to the AS fieldwork component. Therefore teachers can largely plan any investigation/s and thereby ensure that all requirements are met. The teacher can plan, monitor and control much of the work completed in the field and in class. Equally, students can also complete work out of the classroom without supervision. Students can also undertake fieldwork in groups. In all instances, the Head of Centre will have to endorse that the fieldwork has been undertaken.

The examination

There is no requirement for examination candidates to provide a written report of their fieldwork for this assessment component. Fieldwork will be externally assessed

Knowledge check 3

Consider what type/s of statistical technique/s you would use for the following scenarios:

- changing pebble sizes from one end of a beach to another
- examining the differences in the distribution of different ethnic groups within wards in a city
- how the concentration of PM_{10} particles changes with distance from the centre of an urban area
- examining the varying orientation of the long axes of drumlins in two areas of study

in the compulsory Section B of AS Paper 2, where a series of structured questions will be set. The total number of marks is 40 (50% of the whole paper). Questions will be set on materials that are both 'familiar' to the student and 'unfamiliar'. The former will be on general field techniques and the fieldwork the student has undertaken on the 2 days. The latter will be based on new data provided to the student on the day of the examination. Students will be expected to process these data, which may include presentation, analysis and evaluation.

For AS and A-level geography, all assessments will test one or more of the following Assessment Objectives (AOs):

- **AO1:** Demonstrate knowledge and understanding of places, environments, concepts, processes, interactions and change, at a variety of scales.
- **AO2:** Apply knowledge and understanding in different contexts to interpret, analyse and evaluate geographical information and issues.
- **AO3:** Use a variety of relevant quantitative, qualitative and fieldwork skills to: investigate geographical questions and issues, interpret, analyse and evaluate data and evidence, construct arguments and draw conclusions.

Exam tip

Keep a field notebook of your 'familiar' fieldwork. You can use it as a key revision tool before the exam, as it will help you to remember the sites you visited and exactly what you did, and how and why you performed particular activities.

General advice on the examination

The nature of the examination questions is different from that of previous years, and takes the following forms:

(a) Some of the marks are allocated to questions that assess fieldwork techniques in general, such as sampling methods and methods of data collection, presentation and analysis. These combine a mixture of AO1, where knowledge of the chosen techniques is required, and AO2, where evaluation of those techniques is to the fore.

(b) Some of the marks are allocated to questions that invite consideration of a novel situation where the above techniques can be utilised. This requires you to consider fieldwork situations that you may not have encountered. These questions assess AO2. Examples of this include being given a map, or an aerial photograph, or a ground level photograph of an area, and being invited to consider what fieldwork activities could take place in that area.

(c) Some of the marks are allocated to the actual fieldwork experience that you have encountered in your 2 days out of the classroom, although the total number of marks for this is restricted. Hence it is suggested that you experience a complete fieldwork investigation, but that it is limited in scope. You should also reflect on how that piece of fieldwork could be further developed or enhanced. These questions collectively assess AO1 and AO2.

(d) Approximately one-third of the marks is then allocated to a series of questions where you are provided with data collected by a third party in a fieldwork context. These data may be from a physical or a human context from within the specification. The data is likely to be totally unfamiliar to you — indeed, some have called it 'ambush data'. You can then be asked how you could have collected such data, before being asked to present it, analyse it and come to a conclusion based on either or both of these techniques. This series of questions is testing the application of a range of skills, and hence is assessing AO3.

Exam tip

Prepare a list of the types of questions you could be asked regarding the familiar aspects of your fieldwork. Practise responses using sample exam questions and work with mark schemes.

An enquiry focus

The fieldwork undertaken at AS should enable you to develop an enquiry focus (see Figure 22), including the:

- collection of **primary** and **secondary data**
- use of appropriate forms of data presentation, analysis and interpretation
- drawing of conclusions and evaluation of the whole investigation

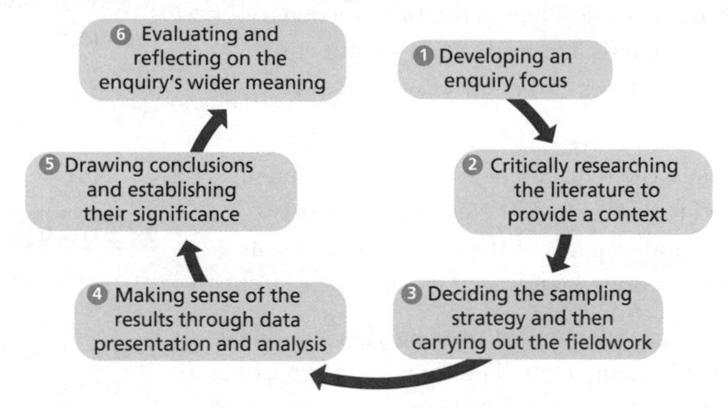

Figure 22 The enquiry process

To prepare for your fieldwork activities, you should begin to pose geographical questions and/or devise suitable hypotheses, consider appropriate primary data collection methodologies and design survey strategies. You should think about selecting appropriate raw data/information (primary data) to be collected in the field, both qualitative and quantitative. This may involve taking measurements and carrying out surveys (for example, questionnaires, observations and interviews), and making images, including field sketches and photographs. You should also consider appropriate sampling techniques. In order to understand the theoretical or comparative context of your research question/s, you should read and/or collect secondary data/information. After your various fieldwork activities, you should then consider appropriate methods of data/information presentation and interpretation, analyse patterns and trends, draw conclusions, and evaluate techniques and results.

Primary data
Information collected by you for the first time through a personal field investigation.

Secondary data
Information derived from published documentary sources already processed, such as processed census data, research papers, textbooks, websites etc.

Fieldwork at A-level

Specification requirements

Students are required to complete a minimum of **4 days** of fieldwork and, as with AS, there must be a combination of both physical and human geography contexts. Part of this fieldwork could be chosen to support geographical understanding in both the physical and human geography aspects of the course. It could also provide an introduction to the process of producing a high-quality geographical enquiry and report. This will form the Non-Examination Assessment (NEA) that carries 20% of the overall assessment.

In preparation for producing the independent fieldwork report, teachers should guide students towards appropriate research areas, types of research questions and methods of appropriate data collection, presentation and analytical techniques. If students have undertaken 2 days of fieldwork at AS, they will need to complete a minimum of

2 further days at A-level. Of course, this is the minimum requirement and the student, or group, could undertake further individual days when collecting data.

The student's independent investigation must incorporate fieldwork data (collected individually or as part of a group) and may include further research and/or secondary data. Where there is group collection of primary data, students should describe their role in this process and indicate the data that are common material.

You must complete the remainder of the process — developing the title, secondary research, and the subsequent interpretation, analysis and evaluation — individually (see Table 9). Students should also seek to make links and connections between their research and relevant geographical theory.

Centres should make time available to explain the requirements of the independent investigation and written report (NEA), and to guide students towards appropriate themes. Students will probably need to revisit all aspects of the enquiry process, as shown in Table 9, and to further integrate their understanding of geographical skills.

As a general guide, the overall time given to this independent investigation should be commensurate with the 20% weighting for this component. The final NEA report has a recommended length of between 3,000 and 4,000 words. There is no assessment of fieldwork skills in the A-level components Papers 1 and 2.

Table 9 Investigation stage and required level of independence

Investigation stage	Level of independence
Exploring the focus	Collaboration allowed
Title of the investigation/Purpose of the investigation	Independent work
Devising methodology and sampling strategies	Collaboration allowed
Primary data collection	Collaboration allowed
Secondary data collection (if relevant)	Independent work
Data presentation	Independent work
Data analysis and interpretation	Independent work
Conclusions and evaluation	Independent work

The rest of this section examines the requirements of the A-level independent investigation (NEA) in more detail, and follows the sequence of the official mark scheme as provided in the specification.

Area 1: introduction and preliminary research

This area is worth **10 marks** and the assessment criteria state that the following will be assessed:

(1) **To define the research questions that underpin field investigations (AO3):** here the required outcome is that the research question/s is effectively identified and is completely referenced to the specification.

Exam tip

In preparation for the AS examination, it is recommended that you undertake some form of 'write-up', although this need not be on an individual basis. Your group could collectively write up your work.

Exam tip

Prepare a cartographic, graphical and statistics revision booklet that justifies as well as describes the various techniques you used on your quantitative data. These techniques are likely to be used in the questions based on 'unfamiliar' fieldwork data.

(2) To research relevant literature sources and understand and write up the theoretical or comparative context for a research question (AO3): here the required outcomes are that the investigation is well supported by the thorough use of relevant literature sources. Also, the theoretical and comparative contexts should be well understood and well stated.

During NEA assessment, the assessor will make a judgement of performance for each criterion, and then apply a 'best fit' judgement for the two criteria and allocate a level for the whole area. Marks will be awarded from within the range of marks provided within that level. There is no weighting of the criteria — they are equal.

What you should do for each of these outcomes

1 To define the research questions that underpin field investigations (AO3)

In order to define the research questions that underpin field investigations, you should ensure that your independent investigation:

- is based on a research question, hypothesis or issue **defined and developed by you individually**
- relates to **any part of the specification content**

You may discuss together with your fellow students, and with your teacher/field study tutor, general ideas and research for appropriate geographical questions. Following this initial stage you must then finalise the focus of your investigation and draft a title (see below for advice). You must do this on your own — **independence is crucial**.

In the candidate record form (CRF) and final written report you must provide a clear justification and contextualisation of how your enquiry will help you to address your title and explore your theme in relation to the chosen geographical location.

Your teacher should review your independent investigation proposal. Within this review your teacher should ensure that the proposed investigation can suitably access the specification requirements and that there is sufficient scope for you to access the full range of marks available for the NEA. Teachers can give approval of each investigation proposal once they ensure that you have independently devised your own hypotheses and/or questions and/or subquestions, even though the title may be the same as/similar to another student's.

Your teacher/field study tutor should advise you on **health and safety** considerations. These are now well established in schools and colleges, but you may wish to adhere to the Risk assessment box below, from the Field Studies Council (FSC).

Exam tip

You are not allowed to choose from a list of titles or investigations provided by your teacher or a field study centre. However, other sources are available.

Risk assessment

Risk assessment is the fundamental tool to ensure safety is effectively managed. The purpose of the risk assessment process is to identify hazards, assess who may be harmed and how, and manage the hazards through safe systems of work. In line with Health and Safety Executive (HSE) guidelines, centres should follow five steps to risk assessment:

1. Identify the hazards.
2. Decide who might be harmed and how.
3. Evaluate the risks and decide on precautions.
4. Record your findings and implement them.
5. Review your assessment and update if necessary.

The likelihood and severity of the hazard/s occurring can be scored numerically (1 equals low, 5 equals high), with resultant risk being assessed as:

- **More than 10:** take immediate action to either remove or control the risk — e.g. take a less risky option or prevent access to the hazard.
- **8–10:** inform people of the risk and look at ways of reducing it.
- **Less than 8:** monitor the situation closely and aim to reduce risk over the longer term.

All significant findings should be documented and periodically updated unless changed circumstances dictate an earlier review.

Risk assessment A structured process of assessing hazards and risks that should be carried out when fieldwork is planned. The process involves identifying the hazards, deciding who might be affected, assessing the likelihood and severity of the risk and taking measures to eliminate or reduce the risk.

Health and safety is important. Prior to any work 'in the field', you need to consider safety. In most school or college situations, a full risk assessment may be needed before any field trips can take place. Any risk assessment first requires identification of actual or potential hazards and then an indication of how these can be overcome or reduced to an acceptable level.

Some safety issues are very obvious, especially when working in exposed and remote physical environments, but it is the slightly less obvious or less extreme risks that you also need to consider. For example, when working in a coastal location the actual risk of drowning is likely to be very small indeed, whereas slipping on rocks and twisting an ankle is much more likely. In an urban environment the dangers of traffic may be obvious but the need to think about how to carry out interviews and avoid the risk of being isolated, or receiving verbal or other abuse, should be important in your planning. **Make sure what you do is safe, and that others know where you will be**. Be fully aware of what you have done to be safe.

Teachers might wish to have discussions with you in relation to the safety of the investigation you will be undertaking. You can create a table of risk — Table 10 is an example of an initial outline.

Table 10 Table of risk — initial outline

Risk/Hazard	Who might be involved	Level of risk	Precautions
Getting caught by waves when measuring beach characteristics	Myself and four other students	1 = Low	Always collect data from at least 5 m above the swash zone

Exam tip

Risk awareness and risk assessment does not form part of the assessment process, but it is essential that you undertake both prior to fieldwork.

Aim, research question, hypothesis or issue?

The **aim** of an investigation is what you are generally trying to achieve in your fieldwork location. This will depend on time, location, environmental conditions, equipment available and risk assessments. For example, your aim might be:

To study the changes in infiltration rates over time in drainage basin X.

Alternatively, you might want to express what you are investigating as a **research question**:

What factors influence infiltration rates in drainage basin X?

Or:

To what extent does distance from the shore affect the vegetation characteristics of a sand dune system?

Or:

How has gentrification changed the character of place D?

For the last of these, you could also think about breaking down the overall question into smaller subquestions. You could go for two or three such subquestions, such as:

1. *What social and demographic changes have taken place in place D in recent years?*
2. *What have been the impacts of these changes on housing and services within place D?*
3. *What are the attitudes of people to the changes that have taken place in place D?*

You may wish to test one or more hypotheses. A **hypothesis** is a statement based on a question that can be either proved or disproved, such as:

- *A number of factors cause flooding to occur at P.*
- *A range of management strategies, both natural and man-made, are used to protect area P from flooding.*
- *Everyone thinks the flood management strategies at P are effective.*

You may wish to **evaluate an issue** in a local area. Examples could include the following.

- *The plan to build a Tesco Local store in place J has created a range of attitudes among local people.*
- *The increase in atmospheric pollution in town K is unpopular and potentially unhealthy.*
- *The proposed development of new housing in the suburban area of L is unnecessary.*

In choosing the **location** for your investigation, consideration should include: accessibility, safety, and availability of appropriate equipment and resources, including data availability and manageability.

If you are designing your own investigation from the very beginning, it may be worth carrying out a simple **feasibility study** or **pilot survey** to make sure that you do not waste your time and that the information you require is available to you. For example, if conducting a questionnaire, you could try out the questions on a small trial group to see if you get the outcomes you want. You can then modify the technique by tweaking the questions.

Exam tip

Note that if you decide to have a series of subquestions, they must be closely tied together into a single theme or focus. Otherwise the investigation could become too large.

2 To research relevant literature sources and understand and write up the theoretical or comparative context for a research question (AO3)

To research relevant literature sources and understand and write up the theoretical or comparative context for a research question, you should ensure that your independent investigation:

- involves research of relevant literature sources
- demonstrates understanding of the theoretical and/or comparative context for that research question, hypothesis or issue
- demonstrates good practice in terms of referencing and using a bibliography system

Teachers should ensure that you have access to relevant literature sources. This can include textbooks, other written material stored within the geography department, and any other external sources such as library and online facilities.

It is important to start collecting resources early, about both the specific area and the general theme chosen. Note that for each of these you should be aware of the concepts or processes (sometimes referred to as the **underpinning theory** or **theoretical context**) that led to the idea for the investigation in the first place. Make sure you have done research into these before beginning your fieldwork. This could involve making use of textbooks, magazines, journals or online materials. You could also make use of local library or archive material. Any work you do here should be clearly and appropriately referenced in the written report. Further guidance is provided in Area 4 (see page 55).

Note also that you must undertake some, or all, of your work outside the classroom in the field, such that **first-hand (primary) data** have to be collected. This may of course include work undertaken at a field study centre or in a work experience setting. You can undertake this work independently or as part of a group. Any data you collect from other written sources is classed as **secondary data**, and you must collect them independently. If your investigation involves some form of comparison with another location, or another area of study, then the **comparative context** (the similarities and differences between the areas of study) must be made clear in your write-up.

Area 2: methods of field investigation

This area is worth **15 marks** and the assessment criteria state that the following will be assessed:

(1) **To observe and record phenomena in the field and devise and justify practical approaches taken in the field including frequency/timing of observation, sampling and data collection approaches (AO3):**

Here the required outcomes are detailed use of a range of appropriate observational, recording and other data collection approaches, including sampling, and thorough and well-reasoned justification of data collection approaches.

(2) To demonstrate practical knowledge and understanding of field methodologies appropriate to the investigation of human and physical processes (AO3):

Here the required outcomes are detailed demonstration of practical knowledge and understanding of field methodologies appropriate to the investigation of human and/or physical processes.

(3) To implement chosen methodologies to collect data/information of good quality and relevant to the topic under investigation (AO3):

Here the required outcomes are detailed implementation of the chosen methodologies to collect data/information of good quality and relevant to the topic under investigation.

During the assessment of the NEA, the assessor will make a judgement of performance for each criterion, and then apply a 'best fit' judgement for the three criteria and allocate a level for the whole area. Marks will be awarded from within the range of marks provided within that level. There is no weighting of the criteria — they are equal.

What you should do for each of these outcomes

These outcomes all concern the methodologies required to **collect data** for the investigation. There are other considerations that are implied when carrying out these methodologies, which include:

- the difference between primary and secondary data
- identification and selection of appropriate physical and human data
- measuring and recording data using different sampling methods
- description and justification of the various data collection methods

Your investigation must incorporate the observation and recording of field data (primary data) that should be of good quality and relevant to the topic under investigation. It should involve justification of the practical approaches adopted in the field, including frequency/timing of observation, sampling strategies used and data collection approaches. Your outcomes should draw on your own research, including your own field data and/or secondary data, and your experience of field methodologies of the investigation of core human and physical processes.

You may collaborate when planning and selecting your methodologies or sampling strategies. Note that one important element of the teacher approval process of your investigation proposal is to ensure that you have made use of appropriate methodology and sampling strategies. Furthermore, if you do not fully justify your methodology and sampling, you may limit your access to marks.

Primary data collection may be carried out individually or in groups, but there must be evidence of your own collection of data in the investigation. There is a strong suggestion throughout the specification as a whole that students make use of both **quantitative** and **qualitative data**.

Secondary data collection (if relevant) must be carried out independently. You should select your secondary sources of data on your own. The use of both quantitative and qualitative data should also be considered here.

Primary data Information collected by you for the first time through a personal field investigation.

Quantitative data Data in numerical form, often involving measurement, which can often be placed into categories and analysed statistically.

Qualitative data Non-numerical data, such as photographs and sketches, and may involve the collection of opinions, perspectives and feelings from questionnaires and interviews.

Secondary data Information derived from published documentary sources already processed, such as processed census data, research papers, textbooks, websites etc.

Sampling

Prior to the investigation, you should make yourself aware of various forms of sampling, the most common ones being:

- **random sampling**
- **systematic sampling**
- **stratified sampling**

Of course, any investigation should consider using a range of sampling strategies, depending on the circumstances of the investigation.

Random means 'due to' or 'of chance'. Therefore, no pattern should be detectable in any situation. A random sample is one that shows no bias and in which every member of the population has an equal chance of being interviewed or used. Random samples are usually obtained by using **random number tables**.

Systematic sampling is a method in which the sample is taken in a regular way, i.e. every tenth house, every fifth person, or at grid intersections on a map for an area-based sampling exercise.

For **stratified sampling**, samples are selected according to some known background characteristic in the statistical population. In studying the distribution of land-use types in relation to geology, a stratified sample would select points in proportion to the area covered by each type of geology. If 30% of the area were clay, 25% sandstone, 30% chalk and 15% alluvium, then for a total sample of 200 points, 60 would be selected on clay, 50 on sandstone, 60 on chalk and 30 on alluvium. In this way the sample is **stratified** according to a known factor. In investigating the opinions of the local population on some development, the sample should reflect the various interest groups fairly. For example, the sample should account for the sex-age distribution in the population, details of which can be obtained from the census.

You may need to make further decisions about sampling. Many investigations rely on a representative sample from the parent population. This population may, for example, be pebbles on a beach, trees in a forest or residents in an area. All samples should be proportional to the size of the total population, and so sample size is an important consideration. **Sample size** refers to the number of observations or data points that make up a survey or data set. Very small sample sizes will not reflect the statistical population closely, and so are unreliable and can lead to incorrect interpretations and explanations. Large samples can become unwieldy and difficult to process.

As stated, sampling may be random, systematic, stratified or a combination of these. It is important to be able to justify the decisions you make about sampling. Again, you should consider using a pilot study. Which method of sampling you choose to use depends upon the nature of your investigation. The impression often given is that random sampling is usually best, since it should remove the risk of bias. This is not always the case in geographical investigations, since you will often be looking to recognise some sort of theoretical spatial distribution. This might therefore

Random sample Type of sample that shows no bias or pattern — members of the 'population' have an equal chance of being selected.

Systematic sample Type of sample in which sample members from a larger 'population' are selected according to a random starting point and at a fixed interval.

Stratified sample Type of sample that is drawn from a number of strata (layers) of a total 'population', when subpopulations vary, in order that it should be representative of each subpopulation.

Exam tip

You should not make use of primary or secondary data not collected by you individually or as part of a group. This includes that collected by parents and teachers.

suggest systematic or stratified sampling as being more appropriate. In studying the downstream changes in a river's discharge there are advantages in having a systematic sample (equal spacing along the river) since you might want to demonstrate that downstream changes take place successively. In applying a questionnaire linked to people's opinions about an issue, you may need to give out questionnaires in proportion to the potential numbers of people in each interest group (stratified sampling), so you do not get skewed results on analysis.

Remember: you can sometimes use more than one sampling technique at a time. For example, if you want to have some structure in your sample but are concerned about potential bias, you might then decide to use some form of random sampling grafted on to a systematic or stratified approach.

Key adjectives

The assessment statements (see page 43) use several key adjectives in their explanation of the requirements — practical, appropriate, good quality, relevant — and the key noun, justification.

Practical refers to a methodology being both feasible and doable. Clearly one aspect of this is that the proposed methodology is safe and therefore dependent on a full risk assessment having taken place (see Area 1, page 40).

Collecting inappropriate and/or unnecessary data does not assist you. As seen in Area 1, initial planning will have determined a key question or hypothesis based on part of the specification content. Discussions with your teacher and/or fellow students should then focus on the data required to address the key question/hypothesis. This will result in you identifying a range of possible data, some of which may have no relevance to the task. At that point, selection of the most **appropriate** data can take place and methods for collecting these data can then be determined.

The data collected, whatever its format, should be of **good quality** and **relevant** — the greater the amount of such data, the greater the accuracy of the final outcome/s of the investigation. It may not be possible for you to collect large amounts of data, as time will be limited, as will your time (and possibly ability) to process, present and interpret large quantities of data. You must, however, have some understanding of what constitutes an acceptable sample size for your data collection method/s with due consideration given to the accuracy of the data collected. Inevitably, some form of sampling will be required.

Finally, you will be required to describe and provide **justification** for your data collection methods. You must understand how you collected your data. This is likely to be presented as a written step-by-step approach. The probable success of your fieldwork will be greatly enhanced by your ability to be very clear as to how you collected your data. However, a higher-level outcome is your justification of the methods. You should understand why you carried out each aspect of your data collection. By being prepared to be asked, and to answer, evaluative questions on your data collection methodology, you should be able to justify it clearly.

Exam tip

Note that justification concerns not just why you have selected a particular methodology but also why you have rejected other methodologies.

Further points

As stated at the start of this book, there is a strong push for you to use data from online sources. With the increase in availability of such sources of data, much of which has been unprocessed, there is some debate among geographers about whether the traditional distinction between primary and secondary data is out of date.

Some argue that unprocessed data collected by you in this format is a form of primary data — it is only secondary once someone else has processed it. Certainly you could collect crowdsourced data, from an outlet such as Twitter, for the first time. Some have referred to this type of data as 'hybrid data' because:

- the methodology used to collect the opinions is similar to primary data collection
- the data are 'raw' — they have not been collated and analysed before you collect them, again similar to primary data
- the opinions were pre-existing and were not generated as part of your investigation, and so are a secondary source

This is, however, a debate for professional geographers and, whatever the argument, you should make sure that you collect some primary data in a fieldwork context. Note also that you can collect both quantitative and qualitative data within the online context.

There is another important difference between primary and secondary data. When collecting primary data, you control the method of data collection and the data are collected for a specific purpose, for example the hypothesis or question you are investigating. This means that you can identify, account for or correct any methodological mistakes or obvious errors and anomalies.

Secondary data need to be used with care because:

- the data were often collected for a different purpose
- the methodology used may not always be clear
- sample size, and thus reliability, may not be clear
- data may contain errors, and these may not be obvious
- data may be old and out-dated

Secondary data sources are very varied. They can be grouped into three main types: statistical, graphical and written (see Table 11).

Table 11 Types of secondary data sources

Statistical	Graphical	Written
Weather data	Maps and plans	Newspapers
River discharge data	Graphs and charts	Diaries
Census data	Satellite images	Radio, TV
Crime statistics	Photographs	Online sources
Deprivation data	Artistic works	Blogs and social media platforms

A very important part of using secondary research sources effectively is to record information about the source. During research, it is easy to forget to do this (especially when using the internet), but it is important because:

- you may need to find and access the source again in order to check details or get more information
- you may be required to cite sources in the fieldwork report or bibliography

As stated above, you should make sure you use both qualitative and quantitative data. Each of these types of data has its strengths and weaknesses (see Table 12).

Table 12 Strengths and weaknesses of quantitative and qualitative data

Data	Strengths	Weaknesses
Quantitative	■ Precise, numerical ■ Reliable as a result of sampling design ■ Can be analysed statistically ■ Collection can be replicated	■ Poor collection methods can lead to weak conclusions ■ Reduces complex situations and views to numbers ■ Complex analysis can produce simplistic mathematical outcomes
Qualitative	■ People's views and opinions provide a human perspective to numerical data ■ Can suggest new research possibilities based on comments made	■ Can take a long time to collect ■ Analysis can be difficult and outcomes may be tenuous ■ Data are subjective and may not be reliable

Your data collection may involve the use of quite sophisticated equipment, such as flow meters for river studies or callipers for measuring pebbles. Often you can improvise quite easily with floating objects, rulers, ranging poles and measuring tapes. Recognition charts are also readily available for stone roundness and plant identification. Questionnaires are almost always designed for purpose, as are interview questions. You will frequently require data recording sheets, and although some of these are available online, it is often best to design your own for your specific purposes.

It is important to collect data that are both precise and accurate, and hence by definition reliable. The **precision** of a measurement is the degree to which repeated measurements under unchanged conditions show the same results. The **accuracy** of a measurement is how close each measurement comes to the real value. The further a measurement is from its expected value, the less accurate it is.

You could include neat, hand-drawn sketches and maps in this part of the report, as well as labelled photographs in order to show how you used complex equipment and maintained reliability and accuracy. If you are using a questionnaire, consider including an annotated blank form to show the reasons why you chose those questions and why you put them in the order in which they appear.

When you are describing your methodology/ies, consider using tables to summarise your work. Tables 13 and 14 are examples based on collecting primary data and secondary data respectively. Remember: you should also indicate which techniques you carried out on your own and which as part of a group.

Table 13 Primary data collection

Data source	Why used/ Purpose	Method: when/ where	Justification of sampling type (if any)	Problems/ Limitations	Improvements
Field measurements					
Land-use survey					
Field sketch					
Photographs					
Questionnaire/ Interview					

Table 14 Secondary data collection

Data source	Why used/ Purpose	Method: when/ where	Justification of sampling type (if any)	Problems/Limitations	Improvements
Government statistics					
Local area plan					
Local newspaper					
Websites					
Textbook					

Area 3: methods of critical analysis

This area is worth **20 marks** and the assessment criteria state that the following will be assessed:

(1) To demonstrate knowledge and understanding of the techniques appropriate for analysing field data and information, and for representing results, and show ability to select suitable quantitative or qualitative approaches and to apply them (AO3):

Here the required outcomes are effective demonstration of knowledge and understanding of the techniques appropriate for analysing field data and information and for representing results. Also, required is a thorough ability to select suitable quantitative or qualitative approaches and to apply them.

(2) To demonstrate the ability to interrogate and critically examine field data in order to comment on its accuracy and/or the extent to which it is representative, and use the experience to extend geographical understanding (AO3):

Here the required outcomes are a thorough ability to interrogate and critically examine field data in order to comment on its accuracy and/or the extent to which it is representative. There should be complete use of the experience to extend geographical understanding.

(3) To apply existing knowledge, theory and concepts to order and understand field observations (AO2):

Here the required outcomes are effective application of existing knowledge, theory and concepts to order and understand field observations. Note that this is the only place in the assessment of the NEA where AO2 is being assessed.

During the assessment of the NEA, the assessor will make a judgement of performance for each criterion, and then apply a 'best fit' judgement for the three criteria and allocate a level for the whole area. Marks will be awarded from within the range of marks provided within that level. There is no weighting of the criteria — they are equal.

What you should do for each of these outcomes

These outcomes all concern the methodologies required to **present**, **analyse** and **interpret** the field data collected during your investigation. However, note that the emphasis in the first two statements is on *analysis* of the field data, and *interrogation* of that field data. Although you (and possibly your teachers) will consider that presentation is important in this process, it should be noted that presentation is only a means to an end and it is the analysis that should take precedence here.

These two statements are assessing AO3 — the 'Skills' assessment objective. The third statement in this set assesses AO2 — the 'Application' assessment objective. So, it is important that you apply, or link, the outcome of the field observations as determined by the analysis to the underlying knowledge, theory or concepts that were introduced in Area 1 of the assessment process.

Exam tip

Note that *analysis* and *interpretation* are more important than *presentation*.

You should select and use appropriate presentation methods on your own. Similarly, you should select and use appropriate data analysis techniques and independently analyse and interpret the results on your own. You must understand *why* the method/s selected are appropriate for the data concerned, and demonstrate the reasons for their selection and use.

Exam tip

Note that teachers/ field study tutors must not suggest or provide guidance on specific methodologies for you to use.

There are other considerations that are implied when carrying out these tasks. There should be:

- an appreciation of a range of visual, graphical and cartographic methods
- selection and accurate use of appropriate presentation and analytical methods
- description, explanation and/or adaptation of appropriate presentation and analytical methods

When applying these presentation and analytical techniques, it is clear that you should use one (or more) qualitative and one (or more) quantitative technique. Hence you could be expected to use some of the following in your investigation:

- annotation of illustrative and visual material, including base maps, sketch maps, OS maps (at a variety of scales), diagrams, graphs, field sketches, photographs, geospatial, geolocated and digital imagery
- overlays, both physical and electronic
- factual text and discursive/creative material, and coding techniques when analysing text
- numeracy — use of number, measure and measurement
- questionnaire and interview techniques
- atlas maps (only if relevant)

- weather maps, including synoptic charts (only if relevant)
- maps with located proportional symbols
- maps showing movement — flow lines, desire lines and trip lines
- maps showing spatial patterns — choropleth, isoline and dot maps
- line graphs — simple, comparative, compound and divergent
- bar graphs — simple, comparative, compound and divergent
- scattergraphs, and the use of line of best fit
- pie charts and proportional divided circles
- triangular graphs
- graphs with logarithmic scales
- dispersion diagrams
- measures of central tendency — mean, mode, median
- measures of dispersion — dispersion diagram, inter-quartile range and standard deviation
- inferential and relational statistical techniques to include Spearman rank correlation and chi-squared test, and the application of significance tests
- remotely sensed data
- electronic databases
- innovative sources of data, such as crowdsourcing and 'big data'
- ICT to generate evidence of many of the skills provided above, such as producing maps, graphs and statistical calculations

Data presentation and analysis go hand in hand, since the best way to present data is one that encourages or allows analysis and interpretation to take place. The first part of this book illustrates some of the many ways in which data can be accurately and meaningfully presented.

For presentation, bear in mind the following.

- Include a wide range of appropriately chosen representation techniques.
- In a geographical investigation, methods of presenting material spatially (e.g. with maps) will be important. These may be based on existing maps or be specially drawn by you for the purpose in mind.
- Simple techniques often work very well, such as using overlays or using a map as a base upon which to plot other information.
- Decide whether the data need spatial techniques or non-spatial techniques, such as pie graphs.
- Photographs, preferably well labelled or annotated, are almost always useful.
- Computer graphics/geospatial mapping can help, and may be very attractive, but beware that you can overuse them and produce 'death by pie chart'.

The presentation section is best integrated into an analysis or results section. Keep the following in mind.

- Line and scattergraphs are often very powerful aids, and when a trend line or a line of best fit is added they become analytical.
- If you are involved in an investigation that is based on strict 'hypothesis testing' principles, don't forget the traditional and relatively straightforward techniques before you get stuck into deep statistical analysis. Remember: a balance between quantitative and qualitative approaches is often best.

- Statistics should be used with a purpose. Difficulty in drawing a line of best fit on a scattergraph may suggest to you that you should apply a Spearman rank correlation test to see if there is a valid correlation between your data sets.

The first section of this book also provides details on a range of statistical skills including Spearman rank correlation and the chi-squared test. Other statistical tests exist that are not referred to on the AQA specification, such as the Mann-Whitney U Test, or the Student's t-test. You are at liberty to use these if you can, and if they are applicable — it may impress!

Exam tip

Quoting the final outcome of a statistical test is important, but you need to understand what it means in the context of your investigation.

Before using statistical tests, make sure:

- that they are necessary, and that you have chosen the appropriate ones
- you know how to interpret the results of the tests, taking into account the degrees of freedom for your data and significance levels, as appropriate

This part of the assessment is not just about doing and applying the presentation and analytical technique/s. You are also required to independently contextualise and summarise their findings and draw conclusions with regard to the knowledge, theory and concepts that underpin the investigation. This may begin with a descriptive summary of what your processed fieldwork data show. This should then be followed by both explanation and interpretation, during which you offer the meanings emanating from your findings. These will eventually assist you in reaching a conclusion/s for your investigation (see Area 4, below).

Finally, you should always be aware of anomalies in your fieldwork data, which may become more apparent during the presentation and analytical processes. An anomaly might be viewed as being something that deviates from the norm, but for you it might, perhaps, best be described as being 'an irregularity' within any data you collect. It is often difficult to identify anomalies in raw data, unless the anomalies are particularly striking. However, once the data have been presented and processed, any anomaly or anomalies may become immediately apparent. You should then offer descriptions of, and possible explanations for, such irregularities.

Much of your analysis will be in terms of written description and explanation. You need to be clear and precise in your expression, quoting your evidence and justifying your identification of trends, correlations, anomalies and relationships. If done well, your presentation and analysis should run smoothly into your conclusions and evaluation — Area 4.

Area 4: conclusions, evaluation and presentation

This area is worth **15 marks** and the assessment criteria state that the following will be assessed:

(1) To show the ability to write up field results clearly and logically, using a range of presentation methods (AO3):

Here the required outcome is a thorough ability to write up field results clearly and logically, using a range of presentation methods.

(2) To evaluate and reflect on fieldwork investigations, explain how the results relate to the wider context and show an understanding of the ethical dimensions of field research (AO3):

Here the required outcomes are effective evaluation and reflection on the fieldwork investigation. There is a complete explanation of how the results relate to the wider context/s and a thorough understanding of the ethical dimensions of field research.

(3) To demonstrate the ability to write a coherent analysis of fieldwork findings in order to answer a specific geographical question and to do this drawing effectively on evidence and theory to make a well-argued case (AO3):

Here the required outcomes are a thorough ability to write a coherent analysis of fieldwork findings in order to answer a specific geographical question. The student draws effectively on evidence and theory to make a well-argued case.

During the assessment of the NEA, the assessor will make a judgement of performance for each criterion, and then apply a 'best fit' judgement for the three criteria and allocate a level for the whole area. Marks will be awarded from within the range of marks provided within that level. There is no weighting of the criteria — they are equal.

What you should do for each of these outcomes

These outcomes all concern the final stages of the investigation whereby you independently contextualise, analyse and summarise your findings and data, and draw conclusions. You should apply your existing knowledge, theory and concepts to order and understand the field observations, and identify their relation to a wider context. You should write up the field results clearly, logically and coherently using a range of presentation methods and extended writing. You should demonstrate the ability to answer a specific geographical question (or hypothesis) and do this by drawing effectively on evidence and theory to make a well-argued case. The final stages of the investigation require evaluation and reflection on the investigation, including showing an understanding of the ethical dimensions of field research (see below). It is important that you evaluate the findings of your investigation and reach a balanced and supported conclusion on your own.

Another important feature is the final write-up of the report. The specification states the following:

> 'Students are expected to submit a written report which is 3,000–4,000 words in length. This includes all text, text boxes and supplementary material such as photographs and data presentation techniques. It does not include appendices. When attaching appendices students should have examples of raw data only, such as data sheets and questionnaires, rather than every questionnaire completed.
>
> Students who offer work that is below the advisory word count may be penalising themselves by not allowing appropriate coverage of the required assessment objectives. Students who exceed the advisory word count may be penalising themselves through a lack of precision and focus.'

The following steps and questions will help you to conclude and evaluate your fieldwork investigation.

- Look again at your initial aim or aims, commenting on the suitability of your chosen location for what you wanted to do.
- Review and evaluate your methods for collecting both primary and secondary data, pointing to strengths and weaknesses (or limitations).
- Review and evaluate your choice of research question/s or hypothesis/es, discussing their appropriateness in the light of what you accomplished.
- Develop your analysis into broader conclusions linked to geographical theory and/or what you found in your particular location.
- Were your conclusions to be expected, or was there something about your locality that threw up unexpected or unusual results?
- Would you do things differently if you were to start again?
- Suggest other avenues of enquiry that may have arisen.

Ethical dimensions of fieldwork

The **ethical dimensions** of field research are important. You should think about ethical issues that can arise when you work within communities and in natural landscapes. At all times you should be sensitive to, and show consideration for, the human and physical environment, including the other people found there. A simple way to address this is to always follow the Countryside Code or the Scottish Outdoor Access Code when in outdoor areas, and always be polite to people and respect their views, especially if they say they do not want to be involved in your work. If working on a human topic, be aware of any social and/or cultural dimensions that might impact on your work. When filling in questionnaires, always tell people that you are doing research and why you are doing it, and that their views will be expressed anonymously. Gain their consent before proceeding.

Be careful about the underlying direction, or possible bias, of your questionnaire and interview questions — avoid steering the respondent in one direction. It is often better to ask questions where the outcome is 'in favour' or 'against', 'stay' or 'change' than asking a question that has a view in it, and the respondent has to say 'yes' or 'no' to that view. Some researchers say that an alternative involving 'yes' tends to be skewed in that direction as people prefer to say 'yes' than 'no'. If using secondary data from an available source, consider gaining permission to use the material, if possible.

In summary, the ethical dimension of fieldwork can be stated as this: do not leave a 'footprint' of where you have been working, either physically or emotionally.

Writing up the report

Keep all your investigation work together in a separate folder. Organise this into sections for easy retrieval. The final write-up of your investigation should be well structured, logically organised, and clearly and concisely written. There are three aspects of this process that you should consider: structure, language and presentation.

Exam tip

Your teacher/field study tutor must not:

- provide templates or model answers
- provide specific guidance on errors and omissions that limits your opportunities to show initiative yourself
- mark work provisionally and share that mark so that you may then improve it
- give specific guidance on how to make improvements

1 Structure

The structure of the report should help the reader to understand it and should also assist you in organising it logically. The following checklist provides a generalised structure to your report.

- Candidate Record Form (CRF)/Cover sheet
- Title page and contents page
- Executive summary (not a requirement, but advisory)
- The introduction: aims, research questions/hypotheses/issues being examined and scene setting
- Sources of information used
- Methods of data collection
- Data presentation, analysis and interpretation
- Conclusion including overall evaluation
- Appendices and bibliography

You need not write these in the order given — indeed, it may be easier if you do not. For example, the executive summary (if you decide to write one) is perhaps best written at the end of the whole process, as it is only at this stage that the 'whole picture' can be described. The following is a suggested order of completion.

(a) Data presentation, analysis and interpretation

This is the section where you present and analyse your findings. At this stage you will have collected the data, sorted them and selected the most useful pieces. You will know what you have found out, and what it all means. Your results will be complete, and they will be most fresh in your mind at this time. You should be able to interpret each separate section of your results and formulate conclusions for each one. The whole picture may begin to appear in your head.

(b) Sources of information and methods of data collection

Now you can write about what information you collected and the methods you used. Do not forget to discuss any limitations of the methods of collection you used, or also of the data sources themselves.

(c) The conclusion

This should include a summary and an evaluation of all the major findings of your investigation. Do not present anything new to the reader at this stage. Towards the end of this section try to draw together each of the subconclusions from each section of the data analysis into one overall conclusion — the whole picture.

(d) The introduction (aims, research questions/hypotheses/issues and scene setting)

Having written up the bulk of the enquiry, you can now write the introduction, making sure it ties in with what follows. This section is intended to acquaint the reader with the purpose of the enquiry, and the background to it.

(e) Appendices and bibliography

The appendices comprise additional pieces of evidence that may be of interest to the reader but are not essential to the main findings. The bibliography provides detail of the secondary sources that have been used in your research, either as guides or as sources of information. Remember: any diagrams or text you have used or copied from secondary sources must be acknowledged.

For all sources, the following information should be recorded:

(a) Who? (the author/s or the name of the organisation responsible for a website)

(b) When? (date of publication, or if a website, the date the source material was uploaded, or if this is not clear, the date the website was accessed)

(c) What? (the title of a book or article in journals and newspapers, or the main website name)

(d) Where? (the publisher of a book, the journal or newspaper title, or the main website name)

New College Stamford LRC
Drift Road Stamford Lincs.
PE9 1XA
Tel: 01780 484339

(f) Contents page

All the sections of the report should be listed in sequence with accurate page references.

(g) Title page

This states the title of your report. Include also your name, candidate number, centre number and date of completion.

(h) Executive summary

This is not a requirement, but advisory. An executive summary should provide a brief statement (no more than 250 words) covering all the main aspects of the investigation. A good executive summary introduces the subject of the full report, refers to its aims and provides a brief synopsis of the findings. A very good executive report will tempt the reader into reading more by being comprehensible, interesting and stimulating. It should also make sense and read as a separate document from the full report.

2 Language

The quality of language that you use in writing up your investigation is important. You are entirely in control of this aspect of the process, and your style of writing must be appropriate for this exercise. You should avoid poor or inaccurate use of English language. In particular, ensure the following.

- Your sentences should be grammatically correct and well punctuated.
- Your writing should be well structured, with good use of paragraphs.
- Your spelling must be accurate (use a dictionary or your PC/Mac spellcheck).
- You must be clear in your use of specialist terminology and in the expression of your ideas.
- You should be aware that the assessment of your work may be influenced by the above aspects of your writing.

Proofreading is an important part of this process. Prior to submission, make sure you read through the draft from start to finish, and mark any places where there are errors or inconsistencies. If possible, ask someone to do this for you — parents or relatives may help. However, you do need someone who is going to be highly critical of what you have written. A report littered with spelling mistakes or grammatical errors does not impress, and it is likely that your PC/Mac spellcheck may miss something (for example, you may have wished to type 'than' but typed 'that' — spellcheck will recognise this as correct, whereas a proofreader, correctly, would not).

3 Presentation

It is a fact of life that most people are influenced by presentation, and that includes teachers and NEA moderators. Bear in mind the following.

- A neatly presented hand-written or typed/word-processed report is going to create a favourable impression before its contents are read.
- Adequate headings and numbering of pages with carefully produced illustrations will make it easier for the reader to understand what is contained within the report.
- Layout is also important. Do not crowd the pages with dense text, which looks unattractive. Provide adequate margins, use either double or 1.5 line spacing if using a word processing package, and make use of clear heading levels with short paragraphs. More sophisticated reports will number the paragraphs in sequence, although this is not a requirement.
- It is also essential to make sure that you insert maps and diagrams in the correct place in the report — it is irritating to have to flick backwards and forwards when trying to read the document. Make sure all the references in the text are included in the bibliography.
- Reports written within the stated word limit tend to be those that are better planned, structured and executed.
- Make sure you allow enough time to add the finishing touches that give your work the 'final polish'. It goes without saying that this time will be available providing you have not left the completion too near to the final deadline.

You should now be in a position to submit your finished product confident in the belief that it is the best you could have done.

Exam tip

Make sure you get a signed receipt from your teacher that he/she has received your report. Unfortunately, reports can go missing on occasion, so you may need this receipt as insurance.

Exemplar fieldwork contexts

The following sections provide accounts of a series of fieldwork scenarios, each partially written up in the style of an A-level student, at varying levels of ability and in different formats.

Note: the items indicated in **[bold square brackets]** are suggestions for additional items that you may provide as part of your fieldwork — the items themselves are not given here.

None of these examples constitutes a complete investigation, nor are they perfect, but they may provide some ideas, opportunities and discussion points for fieldwork that you could replicate in full or in part. You may want to read them and work out how they could have been improved. They are also intended to provide you with exemplars of the possible sequencing of a fieldwork report. They cover a range of geographical contexts.

Coastal context

Introduction

The aim of this study is to see how beaches vary from one another and to examine what factors may influence the characteristics of a beach. I have chosen to perform my study at some beaches in north Wales because they are the closest beaches to where I live, and they are easily accessible. The sites in north Wales that I am going to get my data from are located approximately 40 miles away from my hometown. I will be looking at three beaches in north Wales. They are located at Rhyl, Abergele and Llandudno.

I will measure data from these three different sites so I can compare them to each other and also try to see if they have any common factors.

[Hand-drawn location map]

Location of the study areas

The first site where I am going to retrieve data from is Rhyl, which is located between Prestatyn and Abergele and about 20 miles from Llandudno. Rhyl beach is a good place to acquire results from because it is a generally flat beach with sand as its main grain material. It is approximately 2 miles long. The beach has had groynes built on it to stop longshore drift and it has also attracted lots of tourists. These are two human factors that may have an influence on the formation of the beach. It has a good range of grain size and is a large and open space, so I can record a good transect on it.

The second site I am going to record data from is at Abergele, which is just 5 miles east of my first site at Rhyl. It is 15 miles away from Llandudno. The beach is quite steep with shingle as its grain material. This is why I have chosen two sites that are close together, to also see how the location of the beach affects its characteristics.

The final site where I am going to get data from is Llandudno beach. While the first two sites are located relatively close to each other, Llandudno is situated much further away. It is 15 miles from Abergele and 20 miles from Rhyl. Llandudno is the most northerly point and is situated between two headlands. This means it will have different waves and wind direction, which will influence its characteristics greatly and will show how much a beach changes from other beaches due to different factors. Llandudno is famous for its tourist attractions and its two wonderful beaches, which tourists visit every year. This must have an impact on the beach and is a human factor, which needs to be considered.

[Photocopied extract of OS maps — one for each beach]

Hypotheses

I decided to test two hypotheses.

1 *If the grain size is large, then the percolation rate of the beach will be high, and if the grain size is small, then the percolation rate will be low.*

2 *If a beach's material is large, then the beach transect should be steep and if the materials are small, then the transect should be gentle.*

Data to be collected

For me to be able to see the changes in a beach I am going to have to collect and/or measure some data from each of the sites. These are:

- grain size
- percolation rates
- a transect of the beach
- the angle of the beach

Equipment

I will use the following equipment:

- tape measure
- ranging poles
- string
- spirit level
- stopwatch
- plastic tubing
- water
- quadrat
- ruler
- grain-size card
- clinometer

Underpinning theory

The definition of a beach is 'a geological formation consisting of loose rock particles, such as sand, gravel, shingle, pebbles, cobbles and even shell fragments along the shoreline of a body of water'. A beach has a dynamic nature — it rapidly responds to any changes in inputs to its system very quickly. These changes could be a human factor. Groynes, for example, may be built on the beach, or a physical change like a seasonal variation may have an impact. Many factors affect the form of a beach — these include wave types, grain size and percolation rates.

There are two types of wave action: constructive waves and destructive waves. Constructive waves move material up a beach while destructive waves move the material down the beach. This is because constructive waves have a strong swash so they can move material up the beach easily, and because it has a weaker backwash the material cannot be moved back down as easily, which means it is deposited and the beach starts to take a flatter shape. A destructive wave, though, is quite the opposite. It has a weak swash but its backwash is strong so it picks up the material at the beginning of the backwash and brings it back down the beach, which again makes the beach's transect different and steeper.

The shape of a beach also depends on whether the material of the beach is sand or shingle or another type of material. On sandy beaches the backwash of the waves removes material easily, forming a gently sloping beach. On shingle beaches the swash is dissipated because the large particle size allows percolation, so the backwash is not very powerful and the beach remains steep. This is because the backwash is weaker, so the wave deposits the sediment it is carrying and is not able to bring it back down the beach, so this forms a steep slope. Also, the actual size of the materials can affect how the beach is formed. If the material particles are small, then the percolation rates are slow, so there will be a strong backwash because not as much water has been percolated. If the size of the particles is large, then percolation rate is very quick and there is a weak backwash.

Methodology

I needed to measure certain data from the beach. I needed to measure the grain size, percolation rate, angle of beach and the transect of the beach. This is the primary data I will be collecting. The secondary data I will need to help me with my hypotheses will be weather maps of the days of the recordings and tide times of the beaches for that day.

I measured this data at Rhyl, Abergele and Llandudno in north Wales. I intended to measure each set of results twice — once in the summer and once in the winter. This was so I could compare the two data results with each other and get a seasonal variation. I could not measure this data by myself because some of the tasks are not possible on my own, so I collected it with two friends.

Primary data collection

Transect

The first task was to establish a line of transect on the beaches. This covered as much of the beach as we could measure at the time because of the tide, or in one case we measured up to a length of 100 m because the tide was that far out. I decided to measure only up to 100 m here because I felt that it would be sufficient to show the shape of the beach. I measured the transect of the beach in 5 m intervals and when there was a great change of height in between the 5 m I made the interval smaller. This was so I could record this change of height and this means I got a more accurate recording of the beach. To actually record the data I needed some equipment. I used two ranging poles to place at the start and end of each 5 m interval. The ranging poles had marks of centimetres going up the side of them, so it was possible to measure the transect of the beach. I also needed a long piece of string to tie around the two ranging poles to create a line between them. This string had knots tied at intervals of 1 m. To make sure that this line was horizontal, I used a small spirit level that I could place on top of the string.

At the beginning of each 5 m interval the piece of string was moved to the middle of the markings on the pole. Then, by adjusting the string on the other pole, I made a horizontal line from one pole to the next by moving it up or down. The spirit level helped validate that the line was horizontal. This allowed me to record the height difference between the two poles. This process was carried out all along the beach until I reached the sea or I reached 100 m.

[Diagram showing two ranging poles on the beach, the string in between, and then an indication to measure height difference of the beach between the two poles]

Percolation rate

My next task was to measure the percolation rate of the beach. This is the rate at which the beach absorbs water. The rate was measured at 10 m intervals along the beach. This should show how the rate differs as you go further along the beach. This is a form of systematic sampling because samples are being taken at regular intervals. I did this measurement by placing a plastic tube into the ground and pouring 500 ml of water into it. It was 500 ml every time to ensure accurate and precise data. As soon as the water touched the ground I started a stopwatch and stopped it when the beach had absorbed all the water. The time was then recorded in seconds. To calculate the percolation rate I divided the volume of the water (500 ml) by the time of infiltration. This process was repeated every 10 m until the end of the transect.

[Labelled photograph of the activity]

Sediment size

One of the main characteristics of the beaches is the sediment size. This influences how much water is percolated into the beach and also has an influence on the transect. To measure the sediment of the beach, I needed quadrats, a ruler and a grain size card. I took samples of the sediment at 10 m intervals along the beach and recorded ten sets of data. Again, this is systematic sampling. I placed the quadrat on the ground, which is a metal grid with 100 squares in it. I decided to take a sample from each small square going diagonal from the top left of the quadrat to the bottom right. This is so I used the same method each time and it also produced a wide range of results. When I chose the sample I measured the length of it with a ruler, or if it was too small to be measured by a ruler, I compared it with the grain size chart. I then recorded the data. At 10 m intervals I also measured the angle of the beach with a clinometer, so I could compare it with the grain size distribution and percolation rate.

Secondary data

Weather maps

For secondary data I retrieved weather maps for the days when I recorded the data and then compared them to each other. I related this to the findings to see if the weather influenced the beach data. If the atmospheric pressure of the day was low then this meant there would be more wind, which creates larger waves. If the atmospheric pressure was high then there would be less wind, which creates calmer waves.

Tide times

I also recorded the tide times on the days when I recorded my data. This was so I could calculate where the tide was going when I went to the beaches so I could get a useful set of data. If I went just when the tide was starting to go out, then the water from the sea would affect the percolation rates and the transect data.

Analysis and interpretation

Hypothesis 1: If the grain size is large then the percolation rate of the beach will be high, and if the grain size is small, then the percolation rate will be low.

[Three scattergraphs: (a) percolation rate vs grain size (winter) — all three sites combined, (b) percolation rate vs grain size (summer) — all three sites combined, (c) a combination of (a) and (b)]

The three scattergraphs show the correlation between the percolation rate and grain size at all three locations in summer, winter, and then taken as a whole. All three of these graphs show a positive correlation between percolation rate and grain size. This means that percolation rate does increase as the grain size becomes larger. This is more evident in the third graph. This means that when grain size is small, less water is being infiltrated into the beach so therefore it produces a stronger backwash because of the large amount of water on the surface of the beach. This would affect the shape of the beach, as a stronger backwash would create a gentler beach. When the grain sizes are large the backwash is weaker because of the large amount of water infiltrating, and this produces a steeper beach.

However, the correlation for each of the winter and summer graphs is not totally clear. A reason for this may be because the tide may have only just gone out for some of the readings so this meant the ground would have been saturated, which would have lowered the percolation rate. Another reason may be because we didn't collect enough data for the summer and winter graphs. Also, in the winter we ran out of time, so we did not have the chance to measure the beach at Llandudno.

To see how strong the correlations were on each graph and to show more evidence to prove whether there is a relationship between percolation rate and grain size I performed a Spearman rank correlation test on the total data set, i.e. both seasons combined.

[Completed Spearman rank table]

The results of this test were sufficient to prove there is a relationship between the two variables. The Spearman rank correlation coefficient (SRC) was 0.881. This means there is a strong positive relationship because it is close to 1. To make sure that the result is significant and did not occur by chance, I tested its significance. The results show that it was 99.99% significant. This means there is a 0.01% chance of it being a fluke. The combination of the scattergraphs (as a whole) and the SRC proved that the first hypothesis could be accepted.

I then compared the two seasons' data using Spearman rank to see if seasonal variation affects the data. From the scattergraphs, there did not appear to be much difference. They both have a slight positive correlation and both had an individual SRC result of 0.50. The results at Rhyl are very hard to compare because the grain size was so small. It is nearly impossible to the human eye to measure the grain size accurately, but generally the results from Rhyl produce very low percolation rates in both winter and summer. Comparison of the seasonal data for Llandudno was not possible for the reasons given earlier — we didn't collect any data in winter.

To conclude this hypothesis, I would say that the results produced from the graphs and the SRC test prove that when the grain sizes are small, the percolation rate is low, and when the grain size is large, the percolation rate is high. I feel this part of the investigation was a success as I made the relevant graphs and tests, which made the relationship more evident and helped me prove my hypothesis.

The things I could have improved for this part of the investigation were that I could have made sure I had enough time to measure all the data in the winter so I could get a wider range of data, which may have given me better results. Another aspect I could have improved was that I could have measured more sites at each beach.

Hypothesis 2: If a beach's material is large, then the beach transect should be steep and if the materials are small, then the transect should be gentle.

The first piece of data I am going to examine is the grain size against the angle of the beach. Generally, as the grain size increases so should the angle of the beach. To try and prove this I created a scattergraph on semi-logarithmic graph paper.

[Scattergraph of grain size vs angle of beach]

I created the graph on semi-log paper so that the grain size is proportionally spread out, because the grain sizes at Rhyl are very small and the pebbles at Abergele are large. The results from this are quite conclusive. There is a positive correlation between the two data sets, with a couple of anomalies. They are both circled on the graph and when not included it makes the correlation even more evident. When these anomalies are not included, the lowest angle with the smallest grain size is 2° and the steepest angle with the largest grain size is 11°. This produces a range of 9°, which implies that the larger the grain size the steeper the beach, which provides evidence to prove my second hypothesis.

I have also provided some transects of the beaches at Rhyl, Abergele and Llandudno with an indication of the grain sizes underneath them. This is to see if there is a clear influence of the grain sizes on the appearance of the beach.

[Three transect diagrams with overlays]

The grain sizes at Abergele and Llandudno are larger than the grain sizes at Rhyl. The reason for this is because the beach material at Rhyl is sand and at Abergele and Llandudno it is shingle. On the transects, this has an obvious influence on the shape of the beach. At Abergele and Llandudno they are both steep beaches that gradually get flatter nearer the sea. At Rhyl the beach is generally flat and its gradient is very low. It is less steep at Rhyl because the grain sizes there are so small they can be compacted very close together, so it creates a flatter beach.

The final part of this section is to compare the different seasons' results to see if the season does have an influence on the results. The transects have the winter data plotted on the base graph paper and the summer results are laid over the top of them on overlay paper, so they are easy to compare against each other. There is a very notable change in the shape of two of the beaches from winter to summer. Winter is steeper and higher. This is only true for Rhyl and Abergele, as I was not able to measure the Llandudno beach in winter.

To conclude, therefore, my hypothesis is correct — if the grain sizes of a beach are large, then the beach is steeper than when the grain sizes are small. The scattergraphs show a positive correlation of this relationship.

Comments on the secondary data

The first difference between the 2 days I measured my data on is that they had different atmospheric pressure. The pressure in the summer was high and the pressure in the winter was low. This means that in the winter it created stronger winds than in the summer. Another thing is that the wind direction is different. In the winter it was a NW wind. This headed straight towards the beaches of north Wales, which again would have created stronger waves.

Conclusion and evaluation

The aim of this study was to see how beaches change from one another and to see what factors influence the characteristics of the beach. In this section I will take all of the results and see if they proved or disproved the aims of my study.

My first hypothesis was to see if larger grain sizes on the beach mean there is a higher percolation rate. I did this because the theory was that if the grain sizes are larger they are not as compact, so more water can infiltrate into the ground at one time, as there are large air spaces between the individual particles. Conversely, when the grain sizes are small they will be compacted very close together, so it will be harder for the water to infiltrate the ground. The results showed a strong positive relationship between grain size and percolation rate. The SRC had a result of 0.881. The size of the grains of the beach determines how large or small the percolation rate is going to be.

The second hypothesis was to see if the grain size affects the shape of the beach. If the beach has larger grains, then the transect should be steep, and if the grain size is small, the transect should be generally flat. By comparing the transects of the beaches together with the grain size it was clear that grain size does have an effect on the gradient of the beach. The grain sizes at Abergele and Llandudno are much larger than the ones at Rhyl, and so is the gradient of the beach. At Rhyl it is generally flat because of the small grain size.

Overall I feel this investigation has been a success. I obtained the data I needed efficiently and was able to get some good results that enabled some meaningful interpretation. This allowed me to prove or disprove my hypotheses and so helped me to complete my enquiry.

There are a few things that could have been improved though. First, I was not able to measure the winter data at Llandudno due to bad time management. So, if I was to perform this study again I would make sure that I collected all the data I needed to get more accurate results. Another improvement I could have made would have been to collect more data from each beach, as the small sample size could have affected the data. I would do this by doing three to five transects on each beach for each of the pieces of data collected.

[Limited bibliography]

Vegetation (sand dune) context

Aim

To investigate the stages of a primary succession and edaphic factors in a sand dune system, based at South Sands, Bridlington.

Hypotheses

1. *The variety of plant species increases across a dune system with distance inland.*
2. *Soil pH will become more acidic across a dune system with distance inland.*
3. *Soil moisture will increase across a dune system with distance inland.*
4. *As distance inland increases, the amount of bare land decreases.*

Underpinning theory

The theory that underpins the aim is as follows.

- Sand dunes are an example of a plant succession, a plant community where the structure of the vegetation develops over time. At each stage, certain species have evolved to exploit the particular conditions present. At first only a small number of species will be capable of surviving the harsh environment.
- These hardy pioneering plants slowly change the conditions by altering things such as the mineral content and the moisture content of the soil.
- As each new plant species takes hold, the above process is repeated.
- The changes the plants make allow other species that are better suited to the new conditions to succeed the old species.
- When the succession reaches the point where it is in balance with the climate, a climatic climax is reached.
- A succession that develops on a sand dune is known as a psammosere.

[Location map]

Location of study area and why it was suitable

South Sands, Bridlington, was suitable for my study because the dune transect was of a manageable size, so that I could collect all the data in 1 day. The dune system also contains a variety of vegetation species and dunes, which will give me opportunity to fulfil my aim and relate my findings to the theory.

The site was easily accessible, as there was a car park close by and steps giving access to the dunes (so it was safe). I also gained permission from the local council to conduct the study. Also, nearby there are services provided such as toilets. Table A shows health and safety considerations for the test site.

Table A Health and safety considerations

HAZARD	RISK (low, medium, high)		CONTROL MEASURE
	Likelihood of risk	**Relative impact of risk if occurred**	
Slips, trips and falls	Medium	Low	Need to be careful to look where you are walking. Wear footwear such as Wellington boots with a good grip. Take extra care on the steep dunes at the back of the transect.
Drowning	Low	High	Don't go near the sea or into the sea. Don't even paddle in the sea. Only conduct the survey at low tide or when the tide is going out. Avoid stormy days when there may be danger of large waves.
Dog mess	Medium	Medium	There are lots of dog walkers in the area. Watch where you step and especially where you are putting your bare hands. Wear gloves and wash your hands thoroughly if contact is made with the mess.

Secondary data

One source of secondary information I used for my sand dune study at South Sands, Bridlington, was the 1:25000 Ordnance Survey map. This was fit for purpose as it helped me to accurately locate the study area by using a six-figure grid reference. It was also fit for purpose because it showed me that it was a safe environment for me to carry out my study, because there was a parking area nearby, toilet facilities, refreshments (a public house) and a public telephone box for emergencies in case there wasn't a mobile phone signal. It also showed me that there was access to the sand dune system and there was a wide sandy beach, which was a good source of sand supply for the sand dunes. The 1:25000 map was also fit for purpose as it helped me to plan my route to get to the area. The map also showed me the facilities, such as the caravan site and holiday village, which might be affected if the dune system became unstable due to excessive erosion, perhaps caused by human activity.

[Photocopy of OS map and hand-drawn sketch map of area]

Equipment

I used the following equipment:

- 30 m tape
- clinometer
- ranging poles
- quadrat
- trowel
- labelled plastic bags for soil samples
- plant identification sheet
- soil acidity testing kit
- moisture meter

Hypothesis 1: the variety of plant species increases across a dune system with distance inland.

First, I marked off the straight line transect across the sand dune system from the embryo dunes through to the grey dunes inland using ranging poles, with one at the start of the dune system and one at the back of the system. I then laid the tape measure along the transect line and measured the length of the transect. The transect was 42 m long. The method of data collection I chose was a systematic interrupted belt transect. The reason I chose this was to gain results that would help me to pick out changes in vegetation along the system, as random sampling has more chance of missing out data. I placed a quadrat of 50 cm x 50 cm alongside the tape measure and took readings for vegetation type and percentage every 3 m along the transect. Within each quadrat I estimated the percentage vegetation and noted the type of vegetation. I ensured that the data was as accurate as possible by checking the vegetation types on a plant identification card and trying to estimate percentages as accurately as possible.

[Annotated photographs of the measurement activity]

This method of data collection was good because it showed the difference in vegetation well, and was particularly useful in the large transect that we used and also gave a representative sample along the transect. It also allowed me to concentrate on small species within an individual quadrat, which I may not have seen had I just been looking at the slope as a whole. If I had chosen random sampling I don't know where my sampling points would end up and they could all finish in the yellow dunes, for example, and this would not have given me a transect through the dunes. However, there are also some disadvantages with the method. Because every inch of the slope isn't being examined, whole species that only occur occasionally could be missed altogether and therefore this could affect any conclusions I drew. The sampling technique I chose can be biased because all the species do not have an equal chance of being recorded, depending on where they are in the transect.

Presentation of data/results

I used a series of kite diagrams. This consists of a slope profile of the sand dune system showing the angles and distances measured, drawn to scale on graph paper. The horizontal scale was 1 cm represents 2 m. I placed kite diagrams directly beneath this to see how the vegetation varied along the transect and with slope angle. The kite diagrams show the percentage of each vegetation type found at each sampling point along the transect. Each different type of vegetation can be plotted on a separate kite diagram directly beneath each other to see if certain vegetation types are found in different positions along the transect and on certain slope angles. The x-axes of the kite diagrams are drawn to the same scale as the slope profile (1 cm:2 m), with the y-axis plotted as a percentage with 50% above the x-axis and 50% below the x-axis at a scale of 1 cm:50%. Vegetation is then plotted at the sampling distance along the x-axis (every 3 m). Half of the percentage is plotted above the x-axis and half is plotted below it. The plotted points on both sides of the x-axis are then joined together with a straight line to produce kite/diamond shapes, and these can then be shaded. If only a trace of vegetation was found at a sampling point this was indicated by a letter T at the correct point on the x-axis. By placing kite diagrams for individual species directly below each other, comparison of vegetation types across the dune system can be made.

[Series of kite diagrams]

When I presented my results to show how the vegetation species increased inland I had to first decide which technique to use. I could have drawn pie charts of vegetation at each sampling point. I decided not to do this, as I would not be able to easily see the changes inland, so it was not ideal for me to use. I chose to use kite diagrams as they show spatial changes inland. The difficulties I had with this technique were choosing the correct horizontal scale to use on my graph paper so that I could fit all the points onto a manageable size of paper. I chose 1 cm:2 m. I then needed to decide on a vertical scale for the percentage. I needed a scale that wasn't too exaggerated, so I chose 1 cm:50%. I then had to work out how many kite diagrams I could fit onto one piece of graph paper in order to compare vegetation types.

I also needed to decide which other data to plot on the same piece of graph paper to help with my conclusions. I plotted soil acidity, soil moisture and slope angle using the same horizontal scale. This was done to compare a variety of data in order to assess reasons why certain types of vegetation were growing along the transect.

Results and analysis

On the yellow dune that I found 12 m inland there were very few vegetation types, the two main species being couch grass (70%) and lyme grass (5%). This helps my understanding because it shows that not many vegetation species can survive in the dry environment (soil moisture reading 1 out of 10 on the soil moisture meter). This shows that couch grass and lyme grass are xerophytic and can survive in dry conditions and alkaline conditions (pH 7.5 on the yellow dune). The area is obviously dry because it was mainly sand, and water can percolate through it easily. Further inland the variety of vegetation species increases, for example in the dune slack 18 m inland there is an increase in meadow grass (30%) and soil moisture reading had increased to 2, probably because this was closer to the water table.

Lots of moss appeared between 27 m and 36 m and was greatest at 30 m (100%). This was found on the grey dune, which can retain moisture better (the soil moisture reading on the soil moisture meter increased here to 4) and due to the addition of organic matter from a greater variety of vegetation types, for example meadow grass (10%), lyme grass (20%), dandelion (trace) and couch grass (5%). The soil pH had also decreased to 6.5, which is less alkaline.

Overall there is an increase in the number of species along the transect, but there are some anomalies. For example, in the dune slack 18 m inland there was a decrease in the number of species present and this was due to human interference, as I found a footpath here. Also the textbook theory of a sand dune indicates the presence of embryo dunes and trees, such as oak and Scots pine, beyond the grey dunes. However, at South Sands these were not present due to human interference. Tractors cleaning the beach had removed the embryo dunes and a car park has been built behind the grey dunes.

Statistical analysis

I used the Spearman rank (SR) correlation test to test the hypothesis '*As distance inland increases, the amount of bare ground decreases.*'

This test is used to see if there is a link or correlation between two variables. First, you need to formulate a null hypothesis and a research hypothesis:

- **Null hypothesis:** '*There is no relationship between the amount of bare ground and distance inland.*'
- **Research hypothesis:** '*As distance inland increases, the amount of bare ground decreases.*'

I ranked the two variables (distance inland and percentage of bare ground) from highest to lowest and put them into a table. I ranked the highest value as 1. Tied values received tied ranks, for example values that would have been 5, 6, 7 all became 6 when tied. I then calculated the difference between the ranks (d) and then squared the result (d^2). I then added together the d^2 values (Σ). I then placed the values into a formula.

[Table of completed Spearman rank exercise]

The result gained for my hypothesis was –0.85. To see if this was a significant correlation for 14 sets of data and for the 99% significance level, I needed a SR coefficient value of 0.65 or more. As my value was higher than this, this means that I can reject the null hypothesis and be 99% confident that there is a significant negative correlation between distance inland and bare ground, i.e. the amount of bare ground decreases with distance inland.

However, there are some anomalies, such as in the dune slack where the amount of bare ground actually increases due to footpath erosion.

Evaluation and conclusions

The method we used to estimate the percentage of each species at a particular point was good because it identified major changes and could develop an idea of whether a species was increasing or decreasing with the slope. However, because it was a visual estimate, people may have different views. For example, one person may estimate a different percentage in the same quadrat. To improve this weakness I could make sure that a number of people made the estimation and that they come to a consensus of opinion to avoid the bias of one person, or I could have used a quadrat that was divided into 10% squares to enable even more accurate percentages. It is also possible that I wrongly identified some species, therefore either increasing or decreasing the amount of another species.

My Spearman result of –0.85 showed that bare ground decreased inland. This may not have been the case at every sample point as some areas even close to the back of the dunes may have had only a small amount of vegetation. These areas may have been missed due to my systematic sampling technique. A way to improve this could be to sample at smaller intervals than 3 m to get more coverage of the transect.

One conclusion I came to was that there is a significant negative correlation between distance inland and bare ground, i.e. the amount of bare ground decreases with distance inland. The Spearman result gained for my hypothesis was –0.85. This conclusion is useful to other people's knowledge of the sand dune system, as it showed that as the bare ground decreases inland then the amount of vegetation due to plant succession has increased. Other results back this up — the further inland, the variety of vegetation species increases. This shows that the dune system has become stabilised by natural succession.

These conclusions are of use to people at Bridlington Council, as they need to manage the dune system so that it doesn't get eroded away and destabilised. If the Council allows too much access to walkers, this may cause the dunes to become mobile, as the vegetation is removed and the wind would blow more sand inland, which could have detrimental effects on the tourist industry, such as caravan parks, car parks and other facilities. The Council may decide to fence off areas to prevent erosion from happening.

My conclusions also showed that soil pH became more acid with distance inland, from 7.5 to 6.5 on the pH scale. This may be of use to farmers, as the natural succession of vegetation has altered the soil and made conditions more favourable for other vegetation to grow, and the soil is more stable and there is less chance of soil erosion and possibly more chance of growing crops in the future.

[Bibliography]

Periglacial context

Background

There has been some debate as to the origin of the valley sediments in the Vale of Edale, north Derbyshire. Some suggest they are alluvial deposits from the local rivers. Others suggest they are made up of weathered material that has moved downslope from the local rocks. Some have suggested they are glacial deposits, such as boulder clays. More recent theories suggest that they could be solifluction deposits that accumulated under periglacial conditions. The differing interpretations concerning these deposits have prompted this enquiry into their formation.

The location

The Vale of Edale is an east–west trending valley excavated to a depth of 300–400 m below the level of Kinder Scout (610 m). It displays a series of slope forms generally considered to be characteristic of periglacial areas. The association of tors, scarp edges, terraces, block slopes and valley-fill sediments is particularly evident on the south-facing slopes.

[Photocopied extract of OS map of location]

The infilling in the valley has no distinctive surface morphology — it seems to form a blanket of superficial material over the lower slopes of the vale, masking the underlying geology. Some have suggested that this material indicates deposition of a number of periglacial solifluction lobes that radiated from tributary valleys. The absence of deeply cut valleys on the north-facing slopes suggests that deposition here was more uniform.

Equipment

I used the following equipment during my fieldwork:

- field notebook
- metre ruler
- set of soil sieves
- plastic bags
- trowel
- compass

Methodology

I visited and sampled ten exposures of the valley deposits. This choice was limited by their availability and was thus restricted to two areas where tributary streams of the River Noe are deeply incised into the deposits. On the north-facing slope, to the south of Barber Booth (grid reference SK 114848), a small tributary flowing alongside the road reveals sections of 3 m to 4 m depth, and I sampled four sites. On the south-facing slope, north of Grindsbrook Clough (Grid reference SK 123860), up to 10 m of sediment is exposed in several sections. The deposits are here seen to rest on weathered rock — the Edale shales — and their vertical nature only allowed for detailed examination and analysis at the lowest (2–3 m). This is due to their potential instability, which presented a health and safety risk. Six sites were located here.

I made field notes concerning the nature and layering of the deposits at all field locations. I took samples of 2–4 kg from all sites for laboratory work an ascertaining particle size by using dry sieving methods. At each of the ten sites, I took the orientations of two aspects of the features.

1 The general orientation of the slope at that point using a compass.
2 The *a*-axis orientations of 50 stones, within the range of 20–120 mm, were determined by first clearing the exposure of weathered or slumped material and measuring the stones as they were excavated. Once again, I used a compass. I then plotted these data as two-dimensional, mirror-imaged rose diagrams.

[Collection of mirror-imaged rose diagrams of stone orientation]

A summary of the data I collected is provided in Table A.

Table A Summary of data

Site/ sample	Particle sizes				Average stone orientation	Slope orientation
	% gravel	% sand	% silt	% clay		
BB1	42.4	26.8	17.0	13.8	175–355	10–190
BB2	64.2	17.9	8.7	9.3	8–188	15–195
BB3	30.0	28.7	23.6	17.6	2–182	0–180
BB4	39.7	25.4	18.9	16.0	5–185	8–188
GC1	72.0	19.0	4.2	4.7	30–210	25–205
GC2	58.9	32.5	3.9	4.6	153–333	105–285
GC3	54.1	30.5	7.6	7.8	27–207	20–200
GC4	49.8	35.4	6.0	8.8	25–205	20–200
GC5	52.6	33.4	4.5	9.5	65–245	45–225
GC6	48.2	37.9	5.9	8.0	92–272	95–275
Range	30.0–72.0 = 42	17.9–37.9 = 20	3.9–23.6 = 19.7	4.6–17.6 = 13		
Mean	51.2	28.8	10.0	10.0		
Standard deviation	12.2	6.6	7.1	4.4		

BB = Barber Booth GC = Grindsbrook Clough 1/2/3 etc. = site number

Results

Particle size

I totalled the spread of particle sizes for all of the valley-fill deposits, taken for all of the sediment samples, for each size group, and plotted them onto graph paper. Particle size formed the *x*-axis of the graph, and cumulative total formed the *y*-axis.

[Cumulative graph of % vs particle size]

Gravel and sand are the dominant components of all the size distributions, comprising on average 51% and 29% by weight respectively. Silt and clay are smaller constituents, with 10% by weight each. This wide range of particle sizes is consistent with a poorly sorted deposit. Variability in the samples is, however, illustrated by the wide range of values, especially for the gravels with a range of 42 and a relatively high standard deviation of 12.2 (see Table A).

Size variability is also apparent between the Barber Booth (BB) and the Grindsbrook Clough (GC) samples. When the mean values for each locality are calculated, gravel plus sand comprises 69% (BB) vs 87% (GC). These differences between the north-facing and south-facing locations probably reflect the influence of bedrock. Above GC coarse-grained beds of Kinder Scout grit form the edges of the Kinder plateau, whereas alternating beds of shale and sandstone underlie the north-facing slopes above BB.

Stone orientation

I performed a chi-squared test to see to what extent the samples exhibited an orientation that differed from the random. However, as there was only a small number of potential 'sites' — ten in total — the sample size was inevitably too small for the exercise to be that worthwhile. Nevertheless, I completed the exercise as shown in Table B.

- **Research hypothesis:** *There is a significant trend in the orientation of the stones of Barber Booth and Grindsbrook Clough.*
- **Null hypothesis:** *There is no significant difference between the observed orientation of stones and an expected random orientation.*

Table B Completed exercise for chi-squared test

			A	B	C
Orientation groupings	**Observed (O)**	**Expected (E)**	**O – E**	**(O – E)²**	**(O – E)²/E**
0–45°	6	2.5	3.5	12.25	4.9
46–90°	1	2.5	–1.5	2.25	0.9
91–135°	1	2.5	–1.5	2.25	0.9
136–180°	2	2.5	0.5	0.25	0.1
					$x^2 = 6.8$

Chi-squared value (x^2) = 6.8

I then tested this result for its significance. At 3 degrees of freedom the x^2 result of 6.8 is below both the 0.05% critical value of 7.82 and the 0.01% critical value of 11.34. Therefore we had to reject the hypothesis that the orientation of the stones did not occur by chance and is not randomly orientated. The null hypothesis must be accepted.

Stone orientation vs slope orientation

I then decided to see if there was a strong orientation of the stones in the direction of the surface slope. I undertook a Spearman rank correlation exercise between the average orientation of the elongated stones and the orientation of the slope (see Table C).

- **Research hypothesis:** *There is a significant relationship between stone orientation and slope orientation at the Barber Booth and Grindsbrook Clough sites.*
- **Null hypothesis:** *There is no significant relationship between the two.*

Table C Completed exercise for Spearman rank

Site reference	Average stone orientation (A)	Slope orientation (B)	Rank A	Rank B	Rank difference (d)	Difference squared (d^2)
BB1	175	10	1	8	7	49
BB2	8	15	8	7	1	1
BB3	2	0	10	10	0	0
BB4	5	8	9	9	0	0
GC1	30	25	5	4	1	1
GC2	153	105	2	1	1	1
GC3	27	20	6	5.5	0.5	0.25
GC4	25	20	7	5.5	1.5	2.25
GC5	65	45	4	3	1	1
GC6	92	95	2	2	0	0
					$\Sigma d^2 =$	55.5
					$6\Sigma d^2 =$	333

Rs = 0.66

I then tested this result for its significance. The result of 0.66 is larger than the 0.05% critical value of 0.56 but not the 0.01% critical value of 0.75. Therefore we can reject the null hypothesis and accept the hypothesis that there is a significant relationship between stone orientation and slope orientation.

There is also a small issue with this analysis, and this illustrates a flaw with the Spearman rank test. For site BB1, there appears to be a large discrepancy between the stone orientation (175) and the slope orientation (10) in the relative ranks. However, if we study Table C, we can see that the orientations are quite close — the difference being determined by the way in which the data has been identified, i.e. in relation to degrees on a circular distribution. Perhaps if the data had been ranked in terms of difference from the perpendicular then the correlation would have been much greater.

Evaluation

The sedimentary characteristics of the valley deposits enable certain conclusions to be drawn regarding the conditions under which the material accumulated. The strong preferred orientation of elongated stones in the direction of the surface slope is consistent with materials being deposited by solifluction. Such materials are believed to form as debris flows slowly downslope under saturated conditions in a periglacial environment. The stones align themselves along the line of least resistance to the pressures in the flowing debris. This mechanism is therefore thought to apply to these deposits in the Vale of Edale area. Saturated ground conditions probably prevailed during the summer seasons during the periglacial times after the last Ice Age due to snowmelt on the nearby high ground and the thawing of ground ice. The meltwater would be most likely to remain in the upper soil layers, especially if permafrost was present, thus assisting in the saturated downslope movement.

The unsorted and coarse nature of the sediments and the recognition of other relict periglacial landforms, such as rock edges, tors and the existence of blockfields, in the Vale of Edale area may be taken as additional evidence of periglaciation and hence evidence of solifluction.

Hydrological context: interception rates

Hypotheses

1 *Vegetation species differ in their relative interception rates because of different plant structures, different canopy properties and different leaf properties.*
2 *Different leaf morphologies also influence the interception capacity, since broad leaves are expected to retain more water than thin needles or small leaves.*

Theoretical background

The process of interception is an important link between the relationship of the hydrological cycle and vegetation. Interception losses can be significant and therefore have a severe impact on the water balance. Interception can also play a role when looking at soil erosion, in particular splash erosion. When rainfall travels through the vegetation canopy, the average drop size changes and some drops become larger and some smaller. This change in the size of drops results in different drop impact on soils.

Terminology

- The term **interception** can be described as a process where water (mostly precipitation) is retained in vegetation canopies.
- The term **interception loss** describes the amount of water that is evaporated before it takes part in the land-bound part (throughfall and stemflow) of the hydrological cycle.
- The amount of water that vegetation canopies can retain is named the **interception storage capacity**.
- The term **interception ratio** is determined by the ratio between the interception loss and the precipitation.

In this investigation, the term 'interception' refers to the difference between gross precipitation and net precipitation, where gross precipitation is the amount of water before reaching the canopy and the net precipitation is the amount of water that reaches the soil after travelling through the canopy, i.e. in the following formula (a):

interception = gross precipitation – net precipitation

The amount of rainfall that falls on top of the canopy is referred to as gross precipitation. The net precipitation or the amount of rainfall that is measured on the ground and under the canopy is given by the formula (b):

net precipitation = throughfall + stemflow

Throughfall is determined by measuring the amount of precipitation that reaches the soil under the canopy. Stemflow is the amount of precipitation that reaches the soil via stem and branches from the vegetation, and will be affected by branch orientation and roughness of the bark.

When vegetation becomes more saturated during a rainfall event the interception capacity decreases. The interception capacity also depends on seasonal changes, because the canopy cover changes over seasons. Finally, wind also plays a role in interception capacity variations. Stronger winds result in lower interception capacities because vegetation is less able to hold water in the canopy — it is blown off the leaves.

When we take a more detailed look into the process of interception, evaporation is also an important factor. Water that is intercepted by the canopy is reduced from that canopy because of evaporation. Interception strongly depends on canopy structure and the evaporation rate. Since the rate of evaporation is generally lower than the precipitation rate, the amount of water that vegetation intercepts will increase with increasing precipitation rates.

Species types and properties

In this study, I investigated a number of different vegetation species.

- *Quercus ilex:* large evergreen oak tree that can grow up to 25 m tall. The leaves are leather-like and are variable in form and size, and are mostly elliptical 2–7 cm long, 1–3 cm broad and sometimes with a spiny edge. It grows on all well-drained soils and is abundant in large parts of the study area.
- *Erica arborea:* shrub or small evergreen tree with a height of 1–4 m. It has small leaves 1–3 mm long, with white flowers that are numerous and small.
- *Pinus pinaster:* a medium-size tree, reaching 20–35 m tall and with a trunk diameter of up to 1.5 m. The bark is orange-red, thick and deeply fissured at the base of the trunk, somewhat thinner in the upper crown. The leaves ('needles') are in pairs, stout (2 mm broad), 12–22 cm long, and bluish green to yellowish-green. The cones on the tree are 10–20 cm long and 4–6 cm broad at the base when closed, green at first, ripening glossy red-brown when 24 months old.

[Photographs of the three types of vegetation]

Equipment

I used the following equipment:

- rainfall gauges (including tipping buckets)

I constructed the rainfall gauges from 1.5 l empty supermarket water bottles. All bottles I used were from the same brand, to avoid small variations in size and volume between different brands. I chose these water bottles above traditional gauges due to transportation limitations. I cut off the tops of the bottles and placed these upside down on the lower half of the bottle. I labelled and numbered all bottles.

After each rainfall event, I visited all rainfall measuring plots and measured the amount of throughfall in the gauges and tipping buckets. After that I emptied the gauges and put them back in place.

[Photographs of gauges and tipping buckets in various locations]

Methodologies

Measuring interception

A common method to determine interception in the field is measuring the net precipitation under the canopy. The net precipitation is a combination of stemflow and throughfall. When the gross precipitation is known, formula (a) (defined in the Terminology section) can be used to calculate interception.

Measuring throughfall is done by placing gauges outside and under canopies to measure the amount of water in the gauges after a rainfall event. Note that in this case, stemflow is neglected due to the difficulty of measuring it. The gross precipitation is measured in the open field, away from canopy influences.

(Precipitation data could also be obtained from local weather stations, but this introduces an uncertainty, because precipitation amounts vary strongly over small distances.)

Measuring precipitation

The basis of reliable interception measurements depends on accurate precipitation data. I set up tipping buckets to provide the precipitation data. Tipping buckets register precipitation by recording the number of tips during a rainfall event. Each tip stands for 0.203 mm, which I determined before the fieldwork. By knowing the amount of tips that are recorded, the amount of rainfall can be calculated.

I set up seven measurement locations over the study area. Before placing the tipping buckets, I carefully chose the exact locations. Some tipping buckets had to be placed in an open space to avoid interception of rainfall by trees or shrubs. Practical considerations had to be made to avoid human interference and disturbances, such as walking trails and farming activities. I chose other plot locations based upon vegetation homogeneity and the absence of understory vegetation that might disturb the interception measurements.

The exact location of the measurement plots was however affected by some restrictions. These restrictions were induced by practical and technical limitations. The practical limitations refer to the accessibility of the plots and the time necessary to reach the plots. The technical limitations refer to the availability of free space under the tree and the homogeneous character of the vegetation at the plot. Free space under the tree is required to improve the measurement quality and to avoid interference or interaction with shrubs and grasses. All plots were easily accessed by foot and were within 300 m from a road.

I carried out the rainfall interception measurements for three different vegetation species (see above). After I had chosen the plot locations, I placed five or six (depending on the plot size) rainfall gauges randomly under the canopy. I placed one gauge outside the reach of the canopy and used it to measure the amount of precipitation close to the plot location. I used the data I obtained from the rainfall gauge outside of the canopy to compare with the precipitation data from the tipping buckets under the canopy.

Results

The interception for all three tree species was calculated for every rainfall event using the formula:

interception = precipitation – measured throughfall

To compare the interception for different rainfall events, interception is converted as a fraction of precipitation:

interception ratio = interception/total precipitation

I constructed a plot locations map using ArcGIS. I put the coordinates of all the plot locations into a GPS. I did this as part of the preparatory work in advance of the data collection. I also recorded the weather conditions, coordinates and general plot properties, such as the type of forest and the local topography. I plotted all data including precipitation rates and interception rates onto this map.

[Printout of various elements of the ArcGIS map]

Conclusion

In this study, fieldwork data provided a detailed quantitative interception map. For the investigated sites, I derived interception losses during field experiments by measuring throughfall. The average interception rate for all species was 40%. However, this interception rate may be overestimated because of the limited number of rainfall gauges. Also, some species had only few observations, making them less reliable for statistical analysis. The deciduous and evergreen species have higher interception values compared with the pine species. These results show that for the studied vegetation species, different interception rates are measured.

However, interception values do not only vary per vegetation species — interception values also depend on the rainfall event characteristics, causing different interception losses for different events. In general, for all species, I found higher relative interception percentages for smaller rainfall events and lower interception values coincided with larger rainfall events.

Sources of uncertainty in the field experiments concerned the location of the rainfall gauges, possibly preferential throughfall caused by gaps in the canopy structure and the fact that I neglected stemflow in the interception calculation.

When looking at the interception rates for single species only, interception is highest for large-leaved species such as *Quercus ilex* and smallest for bush-like vegetation species as *Erica arborea*. It is clear, therefore, that vegetation species have an influence on the total amount of interception.

Changing places context

Aim

To examine the character of a place (in this case, Burngreave, a district of Sheffield) and what people who live there feel about it.

Background

Burngreave is a district of Sheffield lying just north of the city centre. It is one of the oldest districts of Sheffield, and was mostly rural in the 1700s. However, by the late 1800s Burngreave was impacted greatly by industrialisation and was a home for several heavy steel and engineering industries. Despite these impacts, Burngreave initially remained a home for wealthier residents, the industrialists and professional classes of the city. Over time, the wealthier residents moved out to cleaner areas to the west of the city, such as Dore and Totley, and other social classes moved in.

During the 1950s and 1960s there was an inflow of immigrants from ex-colonies of Britain, such as India and Pakistan, who settled in the area. This trend has continued — today Burngreave is a multicultural community and is one of the most ethnically diverse neighbourhoods in Sheffield. It has become a home to refugees from countries like Somalia, Eritrea, Iraq, Sudan, and for many immigrants from the Caribbean, Pakistan and Yemen, and more recently to economic immigrants from eastern European countries. This cultural richness is reflected in the number of different languages spoken locally and the variety of cultural activities in the area. But what have been the impacts on the community of Burngreave and what do people think about these impacts?

[Location map of Burngreave]

The following tables are census data extracted from the 2011 census.

Table A Secondary quantitative data: population by age group

Variable	Burngreave	Burngreave (%)	Sheffield	Sheffield (%)
All residents	27,481	100	552,698	100
Under 16	8,058	29	114,246	21
16–24	3,638	13	82,941	15
25–34	4,308	16	74,031	13
35–54	6,576	24	142,579	26
55–64	1,944	7	56,077	10
65–74	1,529	6	44,304	8
75 and over	1,428	5	38,520	7

From Table A, we can see that Burngreave has a relatively young population — 58% of its population is under the age of 35. This compares with the Sheffield figure of 49%. Consequently there is a lower proportion of people over the age of 55 (18%) compared with Sheffield (25%).

Table B Population by gender

Variable	Burngreave	Burngreave (%)	Sheffield	Sheffield (%)
All residents	27,481	100	552,698	100
Male	13,949	51	272,661	49
Female	13,532	49	280,037	51

Table B shows that the gender balance of Burngreave is opposite to that of the city of Sheffield, with a 51%/49% split of male to female.

From Table C we can see that the proportion of White British people is less than half that of the city of Sheffield — only just over a third are classed as White British. The next highest ethnic group is Pakistani, with almost one-quarter of the district being of that ethnicity. Burngreave has a higher proportion of all the other ethnicities (except Indian) than Sheffield, with Black African and Arab being the other most significantly different. The low proportion of Indian people within Burngreave is quite striking.

Table C Population by ethnic group

Variable	Burngreave	Burngreave (%)	Sheffield	Sheffield (%)
All residents	27,481	100	552,698	100
White British	10,468	38	446,837	81
Other White	1,225	4	15,707	3
Mixed race	1,286	5	13,289	2
Indian	463	1	5,868	1
Pakistani	6,256	23	21,990	4
Bangladeshi	157	–	3,326	–
Chinese	139	–	7,398	1
Other Asian	764	3	5,803	1
Black African	2,183	8	11,543	2
Black Caribbean	990	4	5,506	1
Other Black	688	3	3,033	1
Arab	2,143	8	8,432	2
Other	719	3	3,966	1

Table D Length of residence in UK

Variable	Burngreave	Burngreave (%)	Sheffield	Sheffield (%)
All residents	27,481	100	552,698	100
Born in UK	18,303	67	488,176	88
Less than 2 years	944	3	12,365	2
2–5 years	1,707	6	11,652	2
Over 5 years but less than 10 years	2,440	9	14,919	3
10 years or more	4,087	15	25,586	5

Table D suggests a high degree of migration into the district with one-third of the population of Burngreave being foreign-born. It must be noted that some of the 'Born in the UK' category may be second-generation migrants. The average for foreign-born within the city is only 12%, which is close to the national average.

Table E Main language for residents 3 years and over (only numbers of 100 or more)

Variable	Burngreave	Burngreave (%)	Sheffield	Sheffield (%)
English	18,834	69	490,407	89
Polish	229	1	2,611	0.5
Slovak	442	2	1,244	0.25
Arabic	1,399	7	5,043	1
Kurdish	201	1	960	–
Farsi	100	0.5	1,017	0.25
Pashto	272	1	807	–
Urdu	1,155	6	4,222	1
Panjabi	789	4	2,743	0.5
Tigrinya	105	0.5	439	–
Somali	786	4	2,074	0.5

A total of 56 different languages were spoken within Burngreave (2011 census).

This information, partially shown in Table E, tells us that only 7 in 10 people over the age of 3 years within Burngreave have English as their main language. A total of 56 languages are spoken within the district, with Arabic, Urdu, Panjabi and Somali being relatively common. This fact will cause some difficulty for public services in the area, such as education, health and social services.

Table F Population by religion

Variable	Burngreave	Burngreave (%)	Sheffield	Sheffield (%)
All residents	27,481	100	552,698	100
Christian	9,550	35	290,299	53
Buddhist	71	–	2,282	–
Hindu	206	–	3,566	–
Jewish	23	–	747	–
Muslim	11,267	41	42,801	8
Sikh	33	–	942	–
Other	69	–	1,961	–
No religion	4,316	16	172,516	31
Religion not stated	1,946	7	37,584	7

Table F shows us that only one-third of the population of Burngreave are Christian (the figure for Sheffield being just over a half). There are more Muslims in Burngreave than Christian. The figure for 'No religion' is half that of Sheffield as a whole, which suggests that religion as a concept is relatively more important in Burngreave than for the city as a whole.

Table G Accommodation type by tenure

Variable	Burngreave	Burngreave (%)	Sheffield	Sheffield (%)
All households	9,906	100	229,928	100
Owned	4,146	42	134,127	58
Shared ownership	64	–	881	–
Social rented	3,866	39	56,917	25
Private rented	1,650	17	35,760	16
Rent free	180	2	2,243	1

As shown in Table G, only two-fifths of people in Burngreave own their own home, compared with three-fifths in the city as a whole. The rented sector is much more important in Burngreave, especially social housing provided mainly by Housing Associations. This information is a key indicator of relative wealth and inequality.

Table H Accommodation type by household

Variable	Burngreave	Burngreave (%)	Sheffield	Sheffield (%)
All households	9,906	100	229,928	100
Detached	750	8	33,219	14
Semi detached	3,142	32	84,575	37
Terraced	3,469	35	62,786	27
Maisonette or apartment	2,523	25	48,635	21
Other	22	–	713	–

As Table H shows, lower-status housing of terraced and maisonette/apartment housing is much more prevalent in Burngreave than in the rest of Sheffield, where just over half are detached or semi-detached. This is another key indicator of income in the data. Incomes in Burngreave are likely to be lower than in Sheffield as a whole.

Table I Car or van availability by household

Variable	Burngreave	Burngreave (%)	Sheffield	Sheffield (%)
All households	9,906	100	229,928	100
No cars or vans	4,861	49	75,968	33
1 car/van	3,766	38	97,721	43
2 cars/vans	1,048	11	45,766	20
3 cars/vans	175	2	8,154	4
4+ cars/vans	56	–	2,319	1

This is another indicator of relative wealth and inequality. As Table I shows, as car ownership is much lower in Burngreave (with only half of the population having access to a car or van compared to two-thirds for the city as a whole), we can again infer from this that income levels are lower in Burngreave.

Table J Economic activity of household reference person

Variable	Burngreave	Burngreave (%)	Sheffield	Sheffield (%)
All households	9,906	100	229,928	100
Economically active and in work	5,153	52	135,474	59
Economically active and out of work	799	8	8,449	4
Economically active student	198	2	6,073	3
Economically inactive — retired	1,926	19	57,161	25
Economically inactive — student	196	2	4,585	2
Economically inactive — carer	476	5	4,578	2
Economically inactive — sick or disabled	802	8	9,979	4
Other	356	4	3,629	2

The data in Table J supports the previous statements regarding wealth, though perhaps not as strongly as might have been expected. There is a slightly lower proportion in work in Burngreave, and twice as many out of work, but the figures for both areas are low. Indeed, there is a lower proportion of people who have retired in Burngreave. This is balanced, however, with higher proportions of people who are carers, or sick or disabled. Each of these three categories tends to have a lower income.

Table K Highest level of qualification

Variable	Burngreave	Burngreave (%)	Sheffield	Sheffield (%)
All residents over 16	19,734	100	452,014	100
No qualifications	6,987	35	109,841	24
Level 1	2,821	14	55,566	12
Level 2	2,585	13	62,097	14
Level 3	1,803	9	71,594	16
Level 4	3,322	17	116,007	26
Other qualifications	2,216	11	36,909	8

Education levels within Burngreave are generally lower across all of the higher levels of education qualification (see Table K). Hence the proportions with the lowest level of qualification, or no qualification, are higher in Burngreave than the city average. Again, such levels of qualification tend to suggest lower-status employment tasks, which are paid lower rates of pay. These data further suggest inequality within the city.

Table L Occupancy — persons per room by household

Variable	Burngreave	Burngreave (%)	Sheffield	Sheffield (%)
All households	9,906	100	229,928	100
Up to 0.5	5,669	57	160,852	70
Over 0.5 and up to 1.0	3,494	35	63,936	28
Over 1 and up to 1.5	504	5	3,700	2
Over 1.5	239	3	1,440	–

Overall, the data in Table K suggests that there is more overcrowding in Burngreave households than the city as a whole. Again, this is an indicator of relative wealth, and the nature of the housing. This links directly to the relatively large proportion of terraced properties and flats within Burngreave.

Table M General health

Variable	Burngreave	Burngreave (%)	Sheffield	Sheffield (%)
All residents	27,481	100	552,698	100
Very good health	11,917	43	258,250	47
Good health	9,310	33	184,343	33
Fair health	4,086	15	75,733	14
Bad health	1,617	6	26,703	5
Very bad health	551	2	7,669	1

The level of general health within Burngreave is not too dissimilar to that of the city of Sheffield overall (see Table M). There are slightly higher figures for the lower levels of general health within Burngreave.

Burngreave and the concept of place

Places have meanings or associations for all people, whether they are residents or outsiders viewing them from a distance. The meaning attached to a place will vary from person to person, and it may well change over time.

In order to explore some of the meanings of Burngreave to its residents today, I searched for symbols or emblems that are used to represent the settlement. However, Burngreave appears to have no common logo or emblem. Christ Church in Pitsmoor Road, which is still prominent today, located within the older part of the district, does not have a logo or image on its publicity material. The logo for the non-denominational Byron Wood Community Primary School is one of two hands (a black hand and a white hand) coming together with the slogan 'Working together'. Perhaps this is a symbol of the multiracial community that exists here?

Primary qualitative data collection, and evaluation

A piece of research I did to explore people's views of Burngreave was to ask residents what symbol they think best represents life in their area today. Would it be the park, the school or perhaps the degree of multiculturalism that exists? However, this was not a successful question, as Burngreave does not possess a dominant feature. I also asked them to give me an opinion of what they felt about Burngreave and whether it has changed. I completed this primary research by conducting a series of interviews on the streets of Burngreave.

[Blank interview sheet of questions asked of residents]

Some of the opinions given during the interviews about Burngreave were rooted in a sense of nostalgia — a yearning or sentimentality for the past and the feeling that the community has seen better days. There was the sense from some that the 'real' or more authentic Burngreave is under threat, through the loss of its community spirit and some amenities. However, as has already been described, Burngreave has no single past — it has changed through time. There is no one historic version of Burngreave.

Others had a more modern image of Burngreave, believing that it is not bound to its past and that change is for the better. They appreciated the existing amenities, but expressed no concern about them altering. The variety of eating places and entertainment venues was seen as a positive because there was something for everyone's tastes. Similarly, these interviewees celebrated the variety of housing available, from old to new, flats to large detached housing, because it meant that there was something to suit everybody's needs and budgets. In short, they liked Burngreave because it offered something for everyone.

Different views were also expressed about Burngreave's connections to other places. Some saw Burngreave as quite distinct from neighbouring suburbs and other parts of the city. Others regarded it as almost seamlessly connected to other places, whether via bus and tram links to other parts of Sheffield or to the rest of the country via the M1, which is reached quickly via the Sheffield Parkway.

However, I also found that, in many circumstances, direct questions from a stranger can be intrusive and inappropriate, particularly when some potential respondents may not speak English as a first language and others may consider themselves too busy to stop and answer my questions. Perhaps my questions were too direct and abrupt to explore the relatively subtle topics such as residents' lived experiences and views of the place in which they live?

So, I gained other sources of evidence for residents' feelings about Burngreave from looking at posts on internet discussion forums. Although they are not a representative sample of the Burngreave population, those who did express a view of Burngreave online were positive about it, generally viewing it as a friendly place to live. People celebrated Burngreave's wide range of local shops and amenities, which include independent stores, several supermarkets, pubs, restaurants, schools, library, a park and small art gallery. People also appreciated its wide range of clubs and societies, including Keep Fit, a choir and a conservation group for Parkwood Springs Wood.

Parkwood Springs is a natural 'wild' space close to the centre of the city of Sheffield, within the district of Burngreave. There is woodland and heath, open parkland, and a range of natural habitats where flowers such as wild lupins bloom, and where rabbits and sparrowhawks can be seen. There are spectacular views over the city centre, other suburbs and the distant moors of the Peak District to the west. The park has a mountain bike trail (MBT), and the conservation group offers bird watching trails and 'bat walks'. **[A video of cycling on the MBT can be observed at:** vimeo.com/52627717**]**

Artistic views of Burngreave

I studied further aspects of Burngreave through artistic connections. A series of photographic panels erected from Spital Hill to Savile Street, a main thoroughfare through Burngreave, has been created. It is entitled 'A common thread', and it seeks to illustrate the positive aspects of multiculturalism in the community. Information about it can be found at: acommonthreadsheffield.blogspot.co.uk

bit.ly/2rgWun7

A local artist, who has lived in the area for over 30 years, has also been inspired by Burngreave and nearby Attercliffe:

bit.ly/2nXkhth

One person's view of life in Burngreave can be heard at:

bit.ly/2o4WbNL

Conclusion

This investigation has shown how a place that seems fairly ordinary can be investigated, with both primary and secondary research. Wherever the place is, it will have a specific past and present that has been shaped by the interaction of global, regional and local processes, including migration. Although 'globalisation' is seen as a current phenomenon, distant influences have shaped places for many years. Places also have meaning for people. Not everyone will see a place in the same way and it is important to explore people's different views. Burngreave provides an interesting place to undertake such research, and I feel that I discovered much about the attitudes of the people who live there, as well as census data.

Some more themes/ideas for fieldwork

Water and carbon cycles

Investigations of rates of infiltration

Infiltration is a key process within a river catchment. Infiltration can be measured using simple infiltration rings, which can be made from plastic piping. Infiltration rates may be affected by a range of factors such as surface vegetation cover, antecedent weather conditions, soil moisture, soil texture, slope and soil compaction, allowing you to conduct a range of related but distinct investigations in a constrained area.

Measurement of the water balance

Catchment discharge is a fundamental parameter in the drainage basin water balance. Use of secondary rainfall and runoff data will allow you to construct water balances for catchments. Measuring stream discharge and rainfall in the field could help you to understand the potential issues associated with these measures. Measurement of rainfall in multiple simple rain gauges will allow you to examine spatial variation in rainfall and its potential impact on creating good estimates of rainfall inputs.

Estimation of carbon stocks in woodland

You can estimate the stock of carbon within woodland. For example, there are standard equations to estimate living biomass of trees from the diameter of the tree measured at a height of 1.3 m. Tree biomass is 50% carbon so it is a simple conversion to work out how much carbon is stored in the tree. Where tree age can also be estimated, either from the girth of the tree, knowledge of the site or from tree ring evidence from similar felled trees, then the rate of carbon sequestration as mass of carbon per year can be calculated. Hence, you could estimate both the stock of carbon and the flux within a woodland context.

Estimation of changes in carbon stocks in peatlands

Peatland depth is easily measured by probing the peatland either with a commercially available peat probe or with sections of narrow threaded rod, which can be connected to make a portable probe. You can use multiple probings of depth in an area to estimate peat volume at a site. Peat volume can be converted to organic carbon stocks by knowing typical peat densities (circa 0.1–0.2 gC per cm^3). Where the age of local peatlands is known, you can convert the total carbon stock to an average flux over this time period by dividing the stock by the age to give flux in units of grams of carbon per m^2 per year.

Estimation of fluvial fluxes

Estimating fluxes of materials in rivers involves the measurement of discharge and of the concentration of the material of interest, such as sediment or carbon. Flux is calculated as the product of discharge and concentration.

For example, you might measure river discharge either from a site with known stage–discharge relationships (for example, a weir) or by measurement of velocity and river cross section. You could measure sediment concentration by filtering water samples. The organic component of sediment can be estimated as the fraction lost after an hour in a furnace at 550°C. Carbon content of sediment is typically 50% of organic content. The fluvial particulate carbon concentration (the amount of carbon being transported in the sediment) in grams per m^3 multiplied by discharge in m^3 per second will give a carbon flux in grams per second ($g\ s^{-1}$). The same calculation can be applied to dissolved carbon concentrations estimated by colourimetry. This approach is particularly relevant to peatland streams where 'brown water' is indicative of high dissolved carbon concentrations.

Appropriate equipment will be needed for all these tasks.

Some more ideas

More fieldwork ideas include:

- investigation of throughfall and how it varies through the year, and by vegetation type
- investigation of drainage basin characteristics: land use, vegetation, slope, soil permeability/infiltration and their impact on river discharge
- investigation to compare the characteristics of two drainage basins/catchments
- investigation of river discharge over selected times in a year to look at river regimes in relation to season
- investigation of a minor storm event and its impact on discharge in a small stream catchment
- investigation of flooding recurrence levels and areas of flood risk/vulnerability (perhaps using GIS)
- investigation of the impact of a sustained period of drought on water supply and water use, vegetation, sales of summer products (ice creams, salads) and summer activities
- phenological investigations to look at the impact of climate change on natural and human activities (for example, the appearance of catkins or snowdrops, first and last marking of lambs, putting sheep inside/lambing)
- investigation of the impact of human activity (for example, urbanisation, agriculture and deforestation/afforestation) in drainage basins

Landscape systems

The study of landscape systems is traditionally well suited to fieldwork. A key field skill for physical geographers is observation, to answer questions such as 'Why is this landscape like that?'. The ability to observe landforms in the field, to systematically record those observations and then apply classroom knowledge of the environment and processes in order to explain the formation of the landforms observed is crucial. Producing annotated field sketches is a good way to formalise this process. Annotating photographs in the field using appropriate apps is another option.

The observation element within fieldwork is also a great opportunity to collect data, whether this is till fabric data from drumlins, gravel size data from coastal spits or sand transport data from sand traps on coastal dunes.

Often such data is collected at various points in space, so approaches to mapping data, such as proportional circles for mean grain size along the spit, can be developed. Data that is geolocated when it is collected using GPS can be imported into Google Earth or GIS packages to explore digital approaches to handling geospatial data.

Many schools have access to well-established fieldwork locations that relate to coastal and/or glacial systems. Desert-based fieldwork could be more problematic. However, the A-Level Content Advisory Board (ALCAB) subject content specifically allows for fieldwork that looks at aeolian processes in this country. So, studies of dune form in relation to prevailing winds, or sand transport on coastal dune systems or along beaches are all possibilities. If you are lucky enough to be offered overseas fieldwork, there is potential for such fieldwork in relatively accessible areas of southern Europe.

Hot desert landscapes

Hot desert landscape fieldwork could include:

- investigation of how wind-blown sand moves on a beach/sand dune and the relative importance of the transport processes
- investigation of the shape and characteristics of wind-blown features in a sand dune complex
- investigation to see if the volume of wind-blown (airborne) sand increases on heavily used beaches, where the surface layer is more disturbed
- investigation of how the activities of humans interfere with the nature of the sand of the beach that has settled naturally
- investigation of how sand movement varies inland on a transect from an open beach into an area of sand dunes
- investigation of the characteristics and effectiveness of a strategy designed to address the causes of desertification
- investigation of the characteristics and effectiveness of a strategy designed to manage the consequences of desertification
- investigation of the impact of a micro-finance scheme aimed at addressing the effects of desertification on a local community

Coastal landscapes

Coastal landscape fieldwork could include:

- investigation of wave characteristics (wave height, frequency, wavelength) along a stretch of coast
- investigation of changing erosion and deposition on a stretch of coast before and after a storm to look at the impact of processes on coastal features (possibly using previous field work records)
- investigation of raised beaches to look at their distribution, height and post-glacial modifications

- investigation of coastal erosion features: cliff height and profiles, mapping of incidence of faults, joints and bedding planes to study the distribution of micro features, for example caves, arches and stacks and the relationship between erosional features and geology maps
- investigation of beach profiles: long and cross transects to map changes in beach material, gradient, pebble length and pebble roundness along a transect from low to high tide and across the width of the beach
- investigation of a spit using a range of transects to study shape, size and type of deposits on windward and lee sides
- investigation of sand dunes using transects to show dune topography, plant zonation and succession studying changes in physical features (infiltration, pH, wind speed, percentage of bare ground) and associated changes in biotic characteristics (percentage of plant cover, species diversity, plant height) (a psammosere study)
- investigation of a salt marsh using transects to show salt marsh topography, plant zonation and succession studying changes in physical features (soil type, pH) and associated changes in biotic characteristics (percentage of plant cover, species diversity, plant height) (a halosere study)
- investigation of impact of humans on coastal environments: foot path erosion, trampling of dunes, beach litter
- investigation of a coastal management scheme/s along a stretch of coast threatened by either erosion or flooding to investigate the impact of management structures on sediment transfer, for example groynes
- undertaking a cost–benefit analysis or studying the effectiveness of shoreline management plans

Glacial landscapes

Glacial landscape fieldwork could include:

- investigation of the size (height of back wall etc.), shape, orientation and distribution of corries in a defined area
- investigation of the distribution and characteristic features of a glaciated valley (long and cross sections, occurrence of striations, distribution of erosional and depositional features, post-glacial modifications)
- investigation of the distribution and formation of depositional features (glacial vs fluvio-glacial deposit analysis: size, shape, stratification) in an area of lowland ice sheet glaciation
- investigation of the size, distribution, shape and stoss end orientation of a drumlin swarm ('basket of eggs' topography)
- investigation of scree to measure slope, degree of sorting, mapping of source and extent of scree and vegetation colonisation to assess if scree is an active or fossil feature
- investigation of glacial till: till fabric analysis (situation, orientation and shape) to map provenance and movement of ice in a defined area
- investigation of kettle holes/lakes to investigate succession (a hydrosere study)
- investigation of vegetation succession on moraines (a lithosere study)
- investigation of discharge from meltwater streams in a currently glaciated environment
- survey of glacier mass balance in a currently glaciated environment

Changing places

Within this section of the specification, you will study at least **two places**: the place where you live or study, and one or more contrasting locations. This approach lends itself to both quantitative and qualitative methods of study, and also to fieldwork and individual study. In contrast with the other human geography core theme (Global systems) this theme should be accessible to all individual students within a class seeking to conduct independent fieldwork. Your local environment provides much scope for fieldwork, which can begin with a range of local studies and explorations. You might conduct your fieldwork by walking down local streets where you may observe and record the different local and global connections, and how these are represented and meanings made from them, in your respective local areas.

Some examples of contexts through which you may do this include investigations of 'clone towns', regeneration, how or why some places are rapidly changing compared with others, the changing demographics or social make-up of a place, and connections within and between places. The ways in which different groups of people may experience and perceive places differently may provide ideas for fieldwork. Examples include the distinctive ways in which people with disabilities experience places, and are sometimes excluded from them, whether through the architecture of the place itself or the codes of behaviour and attitudes that prevail there. Local surveys could include how residents understand and see the places in which they live, and how their understandings may sometimes contrast with governmental and corporate representations such as those in place marketing or planning documents.

Some of these ideas could be represented by these questions, all with a focus on fieldwork:

- What's unique here/what's special about your place? You could take photographs and/or make sketches about particular elements of the built environment that help you identify with the place. You might want to talk to people and find out what makes it special to them.
- Who are the people using the spaces? Consider age, gender, religion etc. What is their spatial distribution? This is a study of ethnography. Try to map and record.
- Who is marginalised or excluded from this place? Who is not well represented/found here, e.g. young, old, disabled? Is there a visible 'underclass' (this may include people begging on the street etc.)?
- What does the design and architecture say about the place? How does the age and style of buildings in some way tell us about the character of a place, and how does it influence our opinions, especially as visitors? Also important is an idea of scale and use of different types of building materials, e.g. modern (glass, steel, polished stone) vs traditional (rough stone, brick, concrete).
- What is the 'feel' of the neighbourhood? Use adjectives to describe what your instincts, and those of others, are about particular places. Create a map to show this.
- How could this place be improved? What parts of the urban environment are most disappointing (this could include neglect, dirt, dereliction)? Again, map and provide evidence to support your ideas.
- How is this place different from another area/your other study area? Carry out a comparison of differences: focus on the 'feel', e.g. streetscape, furniture, design, architecture and of course the people.

Some further (more traditional) ideas

Further ideas for changing places investigations include:

- survey of variations in townscape/place landscape
- investigation of changing service provision in villages
- investigation of changes in, or characteristics of, suburbanised villages: population size and structure, employment characteristics, housing and community spirit
- investigation of changes in rural areas associated with rural change: holiday homes, language issues, population size and structure, employment and house prices, and problems of service provision
- investigation of building age, type and quality for possible evidence of gentrification
- investigation of the social characteristics and service structure of inner cities
- investigation of employment changes (quality and number of jobs) in redevelopment areas
- investigation of central areas of a city to look at changes in land use, quality of the environment, footfall and characteristics of cultural quarters
- investigation of student districts in urban areas: population characteristics, service provision, attitudes of local residents and housing quality/tenure
- investigation of variations in ethnicity within urban areas
- investigation of variations in levels of deprivation in urban areas: environmental quality, unemployment rates, crime levels, housing tenure, council tax bands, benefit uptake
- investigation of the environmental quality of various parts of a place
- investigation of the environmental, social and economic impacts of a single, large tertiary employer, e.g. a hospital complex
- investigation of impact of tourism on honeypot sites
- for urban or rural re-branding and/or re-imaging: assessment of the success of flagship projects, e.g. sports sites, festival sites, tourism projects to assess environmental, economic, social and cultural impacts
- for any rebranding/regeneration projects: assessment of their sustainability in terms of linkage and involvement to local community, conflicts, economic success, quality of jobs, impact on poor people in an area and likelihood of being value for money and a permanent success

Global systems and global governance

As stated above, this theme might seem relatively inaccessible to individual students seeking to conduct fieldwork. However, some themes may overlap with Changing places, particularly when looking at the impact of globalisation on a local scale place.

Some ideas include:

- investigation of the impact of migration on a particular community: provision of shops, services, schools, places of worship, distribution of groups, housing types, employment, official services (language), index of segregation
- investigation of the distribution of ethnic food outlets and restaurants in a designated area

- investigation of how people use social networks to maintain contact with families
- investigation of a beach to look at distribution and type of sea-borne materials (after a storm and post clean-up) as well as land-supplied litter and waste
- investigation of water quality and management of water quality in coastal areas (blue flag beaches)
- investigation of threatened coastal environments (e.g. seahorse breeding grounds at Studland)
- investigation of the impact of Chinese or Indian diaspora in a named area
- investigation of the impact of foreign direct investment (FDI) in a named place, e.g. Tata in south Wales/Scunthorpe, Chinese development around Manchester airport, Hinkley Point or Liverpool city centre

Optional areas of study

The optional areas of study may also provide opportunities for fieldwork, for example:

Hazards

- investigation of perceptions of the characteristics and impacts of a hazard (earthquake, volcano, drought, storm)

Ecosystems under stress

- investigation of local nature reserves such as SSSI, RSPB, RAMSAR reserves to research reasons for designation, viability, sustainability issues or on the quality and biodiversity of the reserve
- investigation of Biological Action Plans to assess how successful is the work of local conservation organisations such as wildlife trusts
- investigation of the threats to, and impacts on, ecosystems from tourism
- investigation of impact of trampling on vegetation (percentage of plant cover, species diversity, plant height)
- investigation of succession on the margins of a small lake (a hydrosere study)
- investigation of the effects of burning on heathland/moorland ecosystems
- investigation of woodland management scheme/s: strategies, success criteria and rates
- investigation of conflict/s associated with urban development on fragile environments

Contemporary urban environments

In addition to the place/urban themes/ideas given earlier, you could undertake:

- investigation of the urban form of an area and of how it compares with urban models
- investigation of inequalities within an urban area, spatially and across societies (possible link to cultural diversity)
- investigation of urban microclimate: measuring temperature, relative humidity, precipitation, wind strength, light intensity along a transect from the inner-city to the suburbs recording building height and land-use changes

- investigation of the impact of urban development on hydrology within a named place
- investigation of a sustainable drainage system (SuDS) scheme: rationale, strategies and success
- investigation of an urban river restoration scheme: aims, attitudes, success
- investigation of atmospheric pollution in urban areas: nature, causes, impacts, management and success
- investigation of sustainability within an identified urban area

Resource security

Here, you could undertake:

- investigation of strategies to increase water supply in an area
- investigation of strategies to manage water consumption in an area
- investigation of strategies to manage water quality in an area
- investigation of a conflict over water at a local scale
- investigation of the impact of a thermal power station (oil and coal fired) on local microclimate, water/air pollution levels, transport movements and employment
- investigation of the social, environmental and economic impact of a nuclear power station on a designated area
- investigation of the impact of energy efficiency measures on a named community, to include issues such as recycling, use of solar panels etc.
- investigation of the impact/potential impact of a solar energy farm on a place
- investigation of potential sites for the location of wind farms and/or the impact of existing wind farms
- investigation of coalmining in a former mining area, exploring image, culture, health issues and environment, socioeconomic impact and measures to rebrand
- investigation of a conflict over energy at a local scale

Population and the environment

Several of the population-related investigations within Changing places could also be applicable to this option. However, more health-related investigations could also be carried out such as:

- investigation into the health profile of a town, a district or a community in that town
- investigation into the health provision within a town, district or community
- investigation into food supplies within an area, possibly involving the concept of 'food deserts' (areas that have difficult access to supermarkets, and fresh fruit and vegetables, with an associated high incidence of fast-food outlets)

Other ideas include:

- investigation into any aspect of leisure provision in an area, and the environmental influence of/impact on that provision
- investigation of the environmental impact of a new sports stadium in an area
- investigation into the actual and/or perceived impact of climate change on an area

Summary

After studying this section, you should have an:

- awareness of the specification requirements of the fieldwork elements for both the AS and A-level assessment
- understanding of the nature of the fieldwork that you should undertake prior to the assessment of that fieldwork within the AS examination
- understanding of the demands of the mark scheme for the NEA at A-level, and an awareness of what you could do to satisfy those demands
- understanding of how to structure, conduct and write up the individual investigation that constitutes the NEA
- awareness of the wide range of potential themes and ideas that can be used to formulate the title of your own individual investigation for the NEA

Questions & Answers

AS assessment overview

At AS, **geographical skills** and **techniques** can be assessed in any of the content-based examination questions on any of the papers, and in the specific fieldwork and skills questions in Paper 2 Section B. **Fieldwork** is assessed in the specific fieldwork and skills questions in Paper 2 Section B.

All assessments will test one or more of the following Assessment Objectives (AOs):

- **AO1:** demonstrate knowledge and understanding of places, environments, concepts, processes, interactions and change, at a variety of scales.
- **AO2:** apply knowledge and understanding in different contexts to interpret, analyse and evaluate geographical information and issues.
- **AO3:** use a variety of relevant quantitative, qualitative and fieldwork skills to: investigate geographical questions and issues, interpret, analyse and evaluate data and evidence, construct arguments and draw conclusions.

In this section of the book, five questions are given, following the style of questions used in the AS examination papers. Each question is structured as follows:

- sample question in the style of the examination
- mark scheme in the style of the examination
- examplar student answer/s
- examiner commentary on each of the above

Study carefully the descriptions of the 'levels' given in the mark schemes and understand the requirements (or 'triggers') necessary to move an answer from one level to the level above it. You should also read the commentary with the mark schemes to understand why credit has or has not been awarded. In all cases, actual marks are indicated.

Examination skills

Command words used in the examinations

Command words are the words and phrases used in exams and other assessment tasks that tell students how they should answer the question. Exam papers could use the following command words:

Analyse Break down concepts, information and/or issues to convey an understanding of them by finding connections and causes, and/or effects.

Annotate Add to a diagram, image or graphic a number of words that describe and/or explain features, rather than just identify them (which is labelling).

Assess Consider several options or arguments and weigh them up so as to come to a conclusion about their effectiveness or validity.

Comment on Make a statement that arises from a factual point made — add a view, or an opinion, or an interpretation. In data-stimulus questions, examine the stimulus material provided and then make statements about the material and its content that are relevant, appropriate and geographical, but not directly evident.

Compare Describe the similarities and differences of at least two phenomena.

Contrast Point out the differences between at least two phenomena.

Critically Often occurs before 'Assess' or 'Evaluate', inviting an examination of an issue from the point of view of a critic with a particular focus on the strengths and weaknesses of the points of view being expressed.

Define/What is meant by State the precise meaning of an idea or concept.

Describe Give an account in words of a phenomenon that may be an entity, an event, a feature, a pattern, a distribution or a process. For example, if describing a landform, say what it looks like, give some indication of size or scale, what it is made of and where it is in relation to something else (field relationship).

Discuss Set out both sides of an argument (for and against), and come to a conclusion related to the content and emphasis of the discussion. There should be some evidence of balance, although not necessarily of equal weighting.

Distinguish between Give the meaning of two (or more) phenomena and make it clear how they are different from each other.

Evaluate Consider several options, ideas or arguments and form a view based on evidence about their importance/validity/merit/utility.

Examine Consider carefully and provide a detailed account of the indicated topic.

Explain/Why/Suggest reasons for Set out the causes of a phenomenon and/or the factors that influence its form/nature. This usually requires an understanding of processes.

Interpret Ascribe meaning to geographical information and issues.

Justify Give reasons for the validity of a view or idea or why some action should be undertaken. This might reasonably involve discussing and discounting alternative views or actions.

Outline/Summarise Provide a brief account of relevant information.

To what extent Form and express a view as to the merit or validity of a view or statement after examining the evidence available and/or different sides of an argument.

AS questions

Question 1

Explain the basic principles of sampling when students are carrying out fieldwork to collect data for a geographical investigation. (2 marks)

e Mark scheme: 1 mark per valid point.

Student A

Sampling is undertaken, as in an area there is not enough time to access all of the 'population' of the mass of data that is available a. For example, when conducting a questionnaire you can't ask every person, and in a beach survey you cannot measure every pebble b.

e 2/2 marks awarded. a The candidate provides a valid general statement, b followed by examples. Maximum credit is awarded.

Student B

Sampling is important in geographical fieldwork in order to limit the amount of data that is collected a, but it must also be representative of the whole area/population to be statistically valid b.

e 2/2 marks awarded. a b The candidate provides two valid general statements. Maximum credit is awarded.

Question 2

Study Figure 1, a photograph of an area where a geographical investigation could be undertaken. Using evidence from the photograph, explain why this area is suitable for a variety of geographical investigations. (4 marks)

Figure 1 Photograph of an area for geographical investigation

(e) **Mark scheme: 1 mark per valid point. Allow additional marks for development.**

Student A

I can see a range of environments, physical and human, so I could conduct a variety of investigations in this area [a]. I can see an area of coastline, and a settlement with a large area of car parking, so these would provide the contexts [b]. I could look at the impact of tourism in the area, as I can see a footpath descending to the village in the middle of the photo [c]. I can also see areas of woodland, so I could do some work on interception rates in the water cycle [d].

(e) **4/4 marks awarded.** [a] [b] The candidate provides several valid statements, which include general principles, [c] [d] and some development.

Student B

The area is suitable for a variety of geographical investigations as it is easily accessed. There is a footpath and roads to the village, and I can see some boats in the bay, so you can access the area by sea [a]. This accessibility should make the area reasonably safe to work in, but you should always tell people where you will be [b].

(e) **2/4 marks awarded.** [a] [b] The candidate provides two valid general statements.

Question 3

Using evidence from Figure 1, state two appropriate hypotheses, or questions, for geographical investigation in this area. One should be based on physical geography and the other should be based on human geography. (2 marks)

(e) **Mark scheme: 1 mark per valid hypothesis/question.**

Student A

- Physical: to what extent does the geology that an area has influence the rate of coastal erosion?
- Human: to what extent will tourism change the character of a place?

(e) **2/2 marks awarded.** The candidate provides two valid research questions.

Student B

- Physical: vegetation types will influence the rates of interception and/or infiltration.
- Human: environmental quality varies over short distances in an area.

(e) **2/2 marks awarded.** The candidate provides two valid hypotheses.

Question 4

You have experienced geography fieldwork as part of your course. Use that experience to answer the following questions.

State the aim of your fieldwork investigation.

(a) Explain how the investigation helped you to develop your geographical understanding of the concepts studied. (6 marks)

e Mark scheme:

- **Level 2 (4–6 marks):**
  - **AO1: clear knowledge and understanding of the findings of the study. Appropriate knowledge and understanding of the concepts studied and how this has been developed as a result of the study.**
  - **AO2: clearly applies knowledge and understanding to interpret findings of the study. Interprets findings to suggest how these have supported the wider fieldwork aims by linking directly to improved geographical understanding.**
- **Level 1 (1–3 marks):**
  - **AO1: basic knowledge and understanding of the findings of the study. Appropriate knowledge and understanding of the concepts studied and how this has been developed as a result of the study.**
  - **AO2: basic application of knowledge and understanding to findings of the study. Limited interpretation of findings to suggest how these have supported the wider fieldwork aims. Basic links to improved geographical understanding.**

Note: the following two answers make use of different fieldwork aims.

Student A

(a) Aim: to investigate plant succession on a lithosere in the Padley Gorge area of Derbyshire.

My investigation helped me to understand that succession does not occur in a strict order. Theory states that succession always happens in a clearly defined and highly predictable set of stages called seres, with bare rock dominating the start of a lithosere and deciduous trees dominating the end of the lithosere **a**. However, our fieldwork results showed that in the Padley Gorge lithosere, bare rock covered only 10% of site 1 at the start of the lithosere and 80% at site 5 **b**, whereas ferns and shrubs should have dominated this latter site according to the theory. This allowed us to learn that the theory's ideas were based on a theoretical 'perfect' lithosere, which does not exist in reality **c**.

e 2/6 marks awarded. This response is fairly basic in its statements of outcomes of the investigation — so basic in fact that one could challenge the whole basis of the enquiry. However, as this is likely to have been the decision of the supervising teacher, we must only take what the student has written. **a** There is a simple statement of the concepts being studied, **b** together with a straightforward statement of outcome. **c** The answer ends with a basic statement of interpretation. Mid-Level 1 awarded.

Student B

(a) Aim: to compare pollution levels in a rural upland river (Malham Beck) and a rural lowland river (River Cray).

Our investigation was useful in developing my understanding, as we were able to complete our aim of comparing pollution levels in a rural upland river, Malham Beck, and a rural lowland river, the River Cray. This was useful as we could use the results in order to see how pollution levels varied between two different types of rivers. However, our investigation could be argued to be of no use because we rejected our hypothesis that pollution levels would be greater in the River Cray. We rejected it because there were no significant differences in pollution levels between the two rivers **a**.

However, this was useful in the fact it showed us that pollution levels were highly managed in the UK regardless of the river's location. Again, however, it can be argued that our investigation was not useful as it went against our conceptual view that point and non-point sources would impact pollution. For example, at site 2 near Malham village there was a sewage works, a point source, which was expected to raise pollution levels due to the sewage waste. However, the levels of pollution did not change significantly **b**.

Our investigation was also successful in the fact that the pollution levels did not change at the pastoral farmland area (a non-point source) of site 3 on the River Cray. This showed us that farmers here used relatively little or no pesticides or fertilisers, which could raise pollution levels in the river **c**.

In evaluation of the above, I conclude the investigation was successful in developing my conceptual understanding of geography. This is because our results showed us that UK rivers were not the best to test because they were highly managed to reduce pollution. As we were able to see that both non-point and point sources didn't have a significant impact on pollution levels, this then suggested that other factors, such as local geology, have to be taken into account **d**.

e **5/6 marks awarded**. We should remember that fieldwork undertaken for AS is likely to be both small-scale and controlled by the teacher. **a** This candidate provides a clear outline of the purpose of the investigation and of the main findings. **b** **c** This is then followed by some further detail behind the findings, referring to two specific locations. **d** The final paragraph provides a clear interpretation of the findings and suggests other factors that could have influenced the results. The answer is not constructed in a sophisticated manner, but the ideas stated are sound. Mid-Level 2 awarded.

(b) Evaluate the success of your fieldwork experience and explain how you would make use of an opportunity to revisit the location to develop your enquiry further. (9 marks)

e **Mark scheme:**

- **Level 3 (7–9 marks):**
  - **AO1: detailed knowledge and understanding of the data collection methods used in the enquiry process.**

- **AO2: detailed evaluation of methods to assess their utility, reliability and outcomes. Rational conclusions reached as to how the work could have been improved and/or taken forward in future, thus developing a new situation and set of circumstances from the original enquiry.**

- **Level 2 (4–6 marks):**
  - **AO1: clear knowledge and understanding of the data collection methods used in the enquiry process.**
  - **AO2: clear evaluation of methods to assess their utility, reliability and outcomes. Partial conclusions reached as to how the work could have been improved and/or taken forward in future, thus developing a new situation and set of circumstances from the original enquiry.**
- **Level 1 (1–3 marks)**
  - **AO1: basic knowledge and understanding of the data collection methods used in the enquiry process.**
  - **AO2: basic evaluation of methods to assess their utility and reliability. Basic conclusions reached as to how the work could have been improved and/or taken forward in future.**

Note: the following two answers make use of different fieldwork aims to those in part (a).

Student A

(b) Aim: to analyse the succession of a sand dune (psammosere) in terms of vegetation density, vegetation height, pH content and gradient/slope.

My overall understanding of the theory of sand dune succession improved enormously. The theory of a psammosere is much clearer to me thanks to all my hypotheses being correct, therefore my investigation was successful. Using the ranging poles, tape measure and clinometers I could state that as you move inland the gradient/slope of the dunes increases, for example by 6 to 8° on the embryo dunes and 10 to 15° on the grey dunes. Vegetation density also increased with distance inland, having recorded data with a quadrat and analysed it through the use of the Spearman rank correlation coefficient. My aim was successful, as I have concluded that vegetation density does increase as you move inland (from 0% cover near the beach to 100% cover on the fixed dunes) **a**.

My aim to understand the geographical concept of a psammosere was further mastered when, thanks to the kite diagrams I drew, I understood the type of vegetation located on a sand dune varied as you move inland — varying from prickly saltwort on the embryo dunes all the way to sea couch and fully grown trees on the fixed dunes. Therefore my investigation once again was successful **b**.

On the other hand I have learned that more data should be collected at each site on the sand dunes to get a better set of data to analyse. I should have also taken more into account the human impacts on the dunes, and hence this is something I could focus on should I decide to revisit the site and develop my enquiry further. I could also adopt a different sampling method if I revisited the site and use more random sampling rather than making it more biased towards my views, for example by throwing the quadrat a few more times at each point and taking more time when measuring the vegetation height **c**.

ⓔ **6/9 marks awarded**. This answer has satisfied all of the requirements of Level 2. **a** **b** The first two paragraphs provide some clear statements of success, with some detail of methodology and outcomes. **c** The final paragraph is a little less clear, yet does recognise one avenue of further development — the role of human impacts — and some aspects of improved methodology. High Level 2 awarded.

Student B

(b) Aim: to investigate whether the process of gentrification is occurring in Brockley, southeast London.

We systematically sampled ten roads in Brockley including Braxfield Road and Dalrymple Road, covering a study area of 1.5 km^2. We used a scoring method consisting of a ranking based system to score every fifth house along each road out of 45. This overall score is calculated by adding up scores of 1–5 for nine different categories, judging features seen on the external view of the property that would indicate whether gentrification was occurring or not. These include categories such as: whether the original sash windows have been restored or broken, whether there has been a loft conversion or no signs of renovation, and whether the paintwork was tidy or fading. These categories relate to the aim of our investigation because high overall scores — for example above 30, such as a property on Foxberry Road which scored 43 — would indicate that gentrification is occurring in Brockley. On the other hand, low scores on properties — such as a house on Comerford Road which scored 18 — would indicate that gentrification is not occurring. Based on overall scores of the properties, an average score for the road can be calculated and compared with other roads and an average score of the entire study area can be produced **a**.

This method was suitable and successful, as systematic sampling reduces the effect of human interference on the results, reducing subjectivity and so giving more reliable results. This method was also successful as it allowed a range of categories to be scored, allowing for simple comparison between properties, combined into data for roads, to look for evidence as to whether the process of gentrification was taking place in Brockley **b**.

However, different groups sampled different roads and so the results may still be subjective and therefore give partial conclusions. Our method of sampling every fifth house meant that we may have missed patterns of gentrification. This could be improved on a revisit to the area by reducing the sample interval and sampling every two houses instead. For example, in our investigation we missed a house on Comerford Road, which sold for £750,000 in June 2015 on RightMove, which probably (given the above average selling price) showed aspects of gentrification **c**.

Furthermore, some properties on Comerford Road were infill housing, which were relatively new compared to the Victorian/Edwardian housing targeted by gentrifiers and so did not show signs of ageing, making them hard to score using our method. On a revisit to the area, we should have identified the houses/buildings that would not fit into the category of suitable for gentrification, and we should have therefore adopted a more stratified sampling method. In addition to this, housing on Beecroft Road had been converted into flats. This was also hard to score using our method as no individual is designated to take care of the external features. Therefore to provide an accurate representation of the intentions of the owners perhaps the interior of the property would have had to be viewed, or their views sampled using a resident's questionnaire. The former is probably impossible, but a revisit to the area may allow the latter to take place **d**.

In summary, to improve this enquiry we would need to have made a more complicated or extended survey with more categories or sophisticated methods of data collection, which would increase the accuracy and reliability of our results **e**

e **9/9 marks awarded.** Within the time constraints of the examination, this is a highly detailed and appropriate answer. **a** The first paragraph provides a detailed account of the methodology used, with some statements of outcome. **b** The second paragraph then links the previous one to the theme of success. **c** The student then moves on to the second element of the question — the opportunity to revisit an area and what could be done and why. **d** The following paragraph is even more detailed in its discussion of a new, or developed, situation, with even a sense of evaluation of new methods. **e** The answer is rounded off with a neat conclusion. High Level 3 awarded.

Note: there is only one student answer provided to the following question, which is based on unfamiliar data and tests a number of fieldwork-related skills.

Question 5

A group of students carried out an investigation into the impact of globalisation on the small town where they lived. The local textile factory (ClothMade) had closed down 18 months previously and production was transferred to the company's factory in Bangladesh. The students' aim was to discover whether people in different parts of the town felt that the closure had made the town a better or a worse place to live. Their hypothesis was *'People are more pleased with the factory closure as distance from the factory site increases'.*

They carried out a questionnaire survey in ten places at varying distances from the old factory site. Table 1 shows the table of data that they produced. It shows responses to the question 'Has the closure of the ClothMade factory made this town a better place to live?'

Table 1 Responses to the question 'Has the closure of the ClothMade factory made this town a better place to live?'

Sample area	Distance from site (km)	% yes
1	4.5	56
2	3.5	38
3	2.5	14
4	1.5	12
5	0.5	14

Sample area	Distance from site (km)	% yes
6	1	47
7	2	53
8	3	58
9	4	61
10	5	70

The sites of their survey are shown on the sketch map, Figure 2.

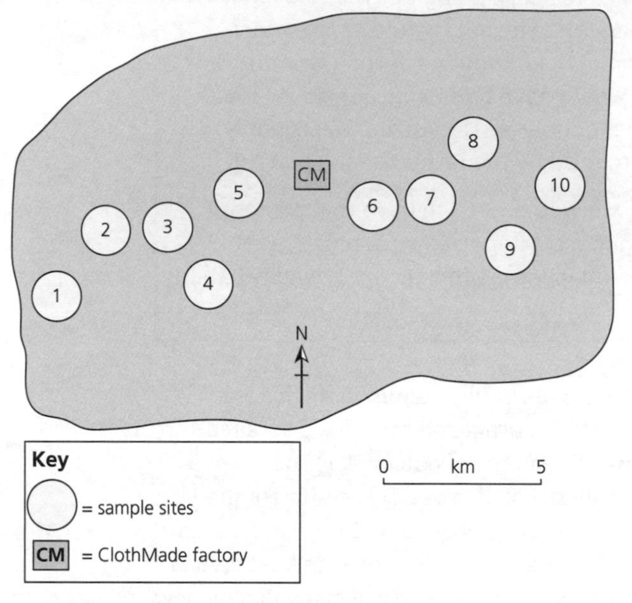

Figure 2 Map of ClothMade factory and survey sites

One of the students tested for a correlation between the two sets of data in Table 1, using a Spearman rank correlation test. Table 2 shows how she set out the data and started her calculations.

Table 2 Student's calculation of the Spearman rank correlation coefficient

Sample area	Distance from site (km)	Rank (distance)	% yes	Rank (% yes)	Difference in rank (d)	d^2
1	4.5	2	56	4	2	4
2	3.5	4	38	7	3	9
3	2.5	6	14	8.5	**2.5**	**6.25**
4	1.5	8	12	10	2	4
5	0.5	10	14	8.5	1.5	2.25
6	1	9	47	6	3	9
7	2	7	53	5	2	4
8	3	5	58	3	2	4
9	4	3	61	2	1	1
10	5	1	70	1	0	0
						Σd^2 = **43.5**

(a) Complete the calculation of Rs (show your working). (4 marks)

e Mark scheme: 1 mark per element of the task. There are four required elements to achieve top marks, shown in bold.

Student answer

(a) The final calculation is as follows:

$$\text{Spearman rank correlation coefficient } (R_s) = 1 - \frac{6 \sum d^2}{n^3 - n}$$

$$R_s = \mathbf{0.736}$$

e 4/4 marks awarded. Required elements are shown in bold in the Student answer and also in Table 2. 1 mark for each number in bold.

(b) Table 3 shows an extract from the table of critical values for Rs. How confident can you be that the student's hypothesis, *'People are more pleased with the factory closure as distance from the factory site increases'*, is supported by the data? (2 marks)

Table 3 Critical values of Rs for Spearman rank correlation coefficient

N	Level of significance (0.05)	Level of significance (0.01)
8	0.643	0.833
9	0.600	0.783
10	0.564	0.746
12	0.506	0.712

e Mark scheme: 1 mark per valid point.

(b) Using Table 3, I can see that there is a 95% certainty that there is a significant correlation (the Rs falls just short of the 99% certainty level), and this correlation has a positive direction **a**. This calculation suggests that people closest to the factory are not pleased to see the factory closed and that those furthest away are **b**. I can therefore be very confident in the hypothesis.

e 2/2 marks awarded. **a b** The candidate provides two valid statements.

(c) The student thought that using a scattergraph to show the data would help her analysis. She drew the graph, shown as Figure 3. Draw a line of best fit on the graph. (2 marks)

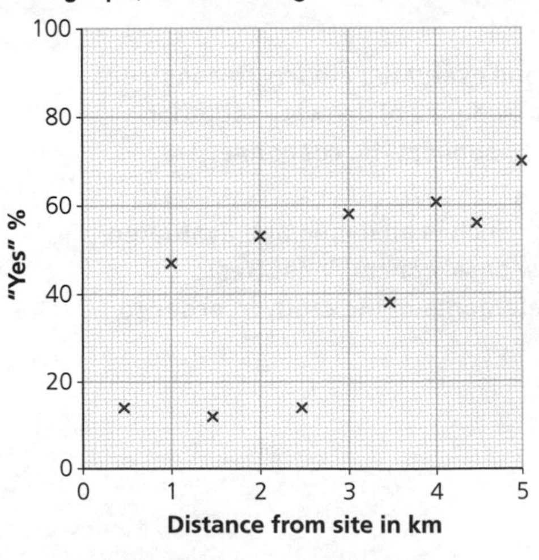

Figure 3 Student's scattergraph

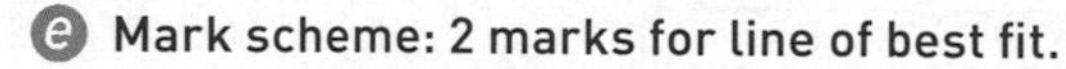

ⓔ **Mark scheme: 2 marks for line of best fit.**

(c)

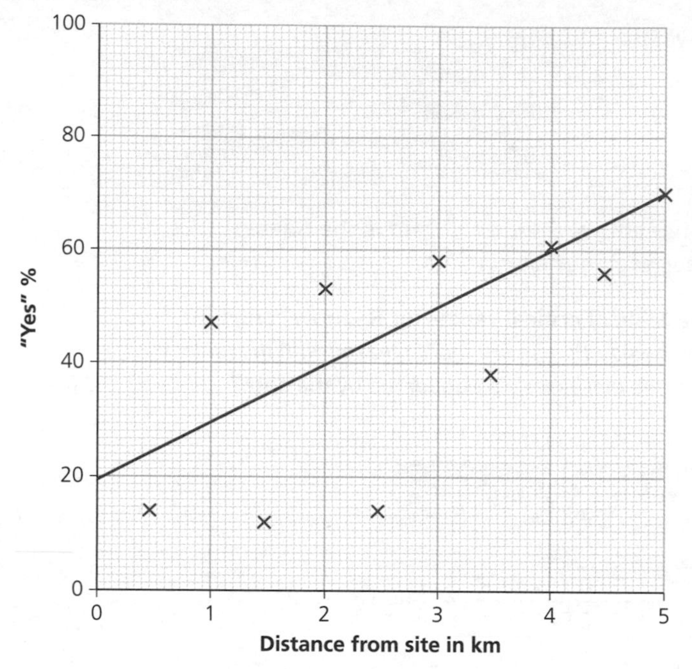

ⓔ **2/2 marks awarded**. The candidate has provided an appropriate line of best fit.

(d) **To what extent does the evidence in this investigation support the hypothesis: *'People are more pleased with the factory closure as distance from the factory site increases'*?** (9 marks)

ⓔ **Mark scheme:**

- **Level 3 (7–9 marks):**
  - **AO3: detailed use of data from the enquiry, which is analysed and evaluated to construct sound arguments and draw valid conclusions. Detailed evidence of drawing together different elements of the study in order to support the response.**
- **Level 2 (4–6 marks):**
  - **AO3: data from the enquiry is analysed and evaluated clearly to construct arguments and draw conclusions. Clear evidence of drawing together different elements of the study in order to support the response.**
- **Level 1 (1–3 marks):**
  - **AO3: basic use of data from the enquiry, which is analysed and evaluated to construct limited arguments and draw basic conclusions. Basic evidence of drawing together different elements of the study in order to support the response.**

(d) From the scattergraph, it can be seen that there is a clear trend supporting the hypothesis, with the line of best fit (LBF) clearly trending from bottom left to top right. However, the correlation is not strong, as very few points are actually close to the LBF. One or more other factors may be involved. For example, if you compare the original data with the map (Figure 2) you can see that people living on one side of the town (1–5) are less pleased with the closure than people living on the other side (6–10), so the hypothesis could be developed further. For instance, it may be that winds from the west blew fumes towards areas 6–10 and so the benefits of closure were greater for people living in that area. On the other hand it may be that the factory drew a higher proportion of workers from housing in areas 1–5, so people in this area were more likely to have lost jobs due to the closure, and so were less pleased.

The outcome of the Spearman rank exercise reinforced the outcome from the scattergraph, as it is clear that the hypothesis could be accepted at the 95% level of significance.

An overall conclusion is therefore that while the linkage between the two variables is clear and that the evidence generally substantiates the hypothesis, close inspection of the data reveals that other factors may be relevant in explaining the apparent relationship, and the anomalies within it. Further investigation could take place.

ⓔ **8/9 marks awarded.** All aspects of Level 3 have been addressed — there is some detailed use of the data, some evaluation and some evidence of synthesis, which includes possible explanation of anomalies within the apparent direct relationship. Furthermore, the student has referred to all elements of the data (Figures 2 and 3 and the Spearman rank coefficient). However, the final sentence could have been elaborated upon. Mid-Level 3 awarded.

Knowledge check answers

1 The type/s of map/s used for the following scenarios would be:
- the population distribution of a country — choropleth or dot map
- the distribution of hospitals in a country — dot map
- the variations in ethnicity in a city — choropleth map
- the origins of customers to a large supermarket — desire line map (if large numbers), trip line map (if small numbers)
- the number of health professionals per 1,000 around the world — choropleth map
- travel times from a major city such as London — a form of isoline map (called an isochrone map)
- the main directions of migrations around the world — flow line or desire line map.

2 The type/s of graph/s used for the following scenarios would be:
- the changing length of a glacier/sand spit/alluvial fan over time — line graph
- the percentage of global deaths from a range of non-communicable diseases — bar graph (simple or compound), or pie graph
- the energy mix of a country — compound bar graph
- the top ten countries with the highest urban populations — bar or pie graphs located on a world map
- the potential relationship between air pollution levels and the number of dementia cases — scattergraph

3 The type/s of statistical technique/s used for the following scenarios would be:
- changing pebble sizes from one end of a beach to another — measures of central tendency, Spearman rank correlation
- examining the differences in the distribution of different ethnic groups within wards in a city — chi-squared test
- how the concentration of PM_{10} particles changes with distance from the centre of an urban area — Spearman rank correlation
- examining the varying orientation of the long axes of drumlins in two areas of study — chi-squared test

Note: **Bold** page numbers indicate key term definitions.

Index

New College Stamford LRC
Drift Road Stamford Lincs.
PE9 1XA
Tel: 01780 484339